Sustainable Development and Green Communication

Also by Jan Servaes

SUSTAINABILITY, PARTICIPATION AND CULTURE IN COMMUNICATION: Theory and Praxis

COMMUNICATION FOR DEVELOPMENT AND SOCIAL CHANGE

MOVING TARGETS: Mapping the Paths between Communication, Technology and Social Change in Communities *(co-edited with Liu Shuang)*

COMUNICACION PARA EL DESAROLLO HUMANO Y EL CAMBIO SOCIAL

INFORMATION SOCIETY OR KNOWLEDGE SOCIETIES? UNESCO in the Smart State *(co-edited with Rhonda Breit)*

TOWARDS A SUSTAINABLE EUROPEAN INFORMATION SOCIETY *(co-edited with Nico Carpentier)*

READ THE CULTURAL OTHER: Forms of Otherness in the Discourses of Hong Kong's Decolonisation *(co-edited with Shi Xu and Manfred Kienpointner)*

COMMUNICATION FOR DEVELOPMENT: One World, Multiple Cultures

THEORETICAL APPROACHES TO PARTICIPATORY COMMUNICATION *(co-edited with Thomas Jacobson)*

MEDIA & POLITICS IN TRANSITION: Cultural identity in the Age of Globalization *(co-edited with Rico Lie)*

PARTICIPATORY COMMUNICATION FOR SOCIAL CHANGE *(co-edited with Thomas Jacobson and Shirley White)*

Sustainable Development and Green Communication

African and Asian Perspectives

Edited by

Jan Servaes
City University of Hong Kong

palgrave
macmillan

First published 2013 by
PALGRAVE MACMILLAN

Palgrave Macmillan in the UK is an imprint of Macmillan Publishers Limited, registered in England, company number 785998, of Houndmills, Basingstoke, Hampshire RG21 6XS.

Palgrave Macmillan in the US is a division of St Martin's Press LLC, 175 Fifth Avenue, New York, NY 10010.

Palgrave Macmillan is the global academic imprint of the above companies and has companies and representatives throughout the world.

Palgrave® and Macmillan® are registered trademarks in the United States, the United Kingdom, Europe and other countries.

ISBN 978–1–137–32940–0

This book is printed on paper suitable for recycling and made from fully managed and sustained forest sources. Logging, pulping and manufacturing processes are expected to conform to the environmental regulations of the country of origin.

A catalogue record for this book is available from the British Library.

A catalog record for this book is available from the Library of Congress.

Contents

Figures and Table

Figures

Table

Contributors

Editor

Jan Servaes is Chair Professor and Head of the Department of Media and Communication at City University of Hong Kong, UNESCO Chair in Communication for Sustainable Social Change and Director of the SBS Center Communication for Sustainable Social Change (CSSC) at the University of Massachusetts Amherst. He is also Honorary Guest Professor at the Huazhong University of Science and Technology (HUST), Wuhan, China, Researcher at the Brussels Center for Journalism Studies (BCJS), Belgium, Editor-in-Chief of *Telematics and Informatics: An Interdisciplinary Journal on the Social Impacts of New Technologies* and of *Communication for Development and Social Change: A Global Journal*. He is Editor of the Southbound Book Series "Communication for Development and Social Change", and, of the Hampton Book Series "Communication, Globalization and Cultural Identity".

He has been President of the European Consortium for Communications Research (ECCR, http://www.ecrea.eu) and Vice-President of the International Association of Media and Communication Research (IAMCR, http://www.iamcr.org), in charge of Academic Publications and Research, from 2000 to 2004. He chaired the Scientific Committee for the World Congress on Communication for Development (Rome, 25–27 October 2006), organized by the World Bank, FAO and the Communication Initiative.

He has taught International Communication and Development Communication in Australia (Brisbane), Belgium (Brussels and Antwerp), the United States (Cornell), the Netherlands (Nijmegen) and Thailand (Thammasat, Bangkok) in addition to several teaching stints at around 100 universities in 43 countries.

He has undertaken research, development and advisory work around the world and is known as the author of journal articles and books on such topics as international and development communication; information and communication technology ICT and media policies; intercultural communication and language; participation and social change; and human rights and conflict management. Recent books include: *Sustainability, Participation and Culture in Communication. Theory and*

Praxis (2013), *Communication for Development and Social Change* (2008), *Moving Targets. Mapping the Paths between Communication, Technology and Social Change in Communities* (2007, co-edited with S. Liu), (2006) Thomas P. & Servaes J. (eds.), *Intellectual Property Rights and Communications in Asia* (2007, co-edited with P. Thomas), *Towards a Sustainable European Information Society* (2006, co-edited with N. Carpentier), (2005) Shi Xu, Kienpointner M. & Servaes J.(eds.), *Read the Cultural Other: Forms of Otherness in the Discourses of Hong Kong's Decolonisation*, (2005, co-edited with Shi Xu and M. Kienpointner), (2003), *The European Information Society: A Reality Check* (2003, ed.), *Approaches to Development: Studies on Communication for Development* (2003, ed.) and *Communication for Development: One World, Multiple Cultures* (2002).

Contributors

Fabien Cishahayo is a consultant and a lecturer at the University of Montreal. He has been teaching communication sciences in numerous Canadian universities (University of Montreal, University of Sherbrooke, University of Ottawa and University of Sudbury). He was Chair of the Public Communication department of the University of Sudbury from 2006 to 2011. His field of specialization is communication for development, especially the uses of (new) media for development purposes. He is also interested in exploring the impact of (ICTs) on diaspora communities. His current research topic deals with the political economy of cloud computing. Dr Cishahayo is involved in various Canadian and international media. Since 2003, he has been a regular contributor to the Voice of America as a political analyst. Also involved in cyberactivism, he launched the blog (http://www.burundibwacu.info) in 2003. In 2008, he published the political essay *"Burundi: Liberté, je crie ton nom"*.

Royal D. Colle is Professor Emeritus at Cornell University, where he has been on the faculty for more than 40 years. He has lived and worked abroad in countries ranging from India and Indonesia to Western Samoa and Guatemala. Colle has served as a consultant for a variety of international organizations including the World Bank, WHO, FAO, the UN, UNFPA, UNESCO, USAID and the Ford Foundation. His work has focused on institution building related to communication; the design of communication strategies for development programs; and innovative uses of information technology for development. His most recent projects have been in Asia and Africa working on university-supported telecenter systems. He is co-author of *A Handbook for Telecenter Staffs* and author of an online book *Advocacy and Interventions, Readings in Development Communication.*

Henri-Count Evans is a climate change communication researcher based in Zimbabwe. He is an MPhil candidate; his thesis is titled "Climate Change Communication: Opportunities and Challenges in Southern Africa". Henri-Count has done research work pertaining to media practice and reporting of climate change issues. The research covered both the print and electronic media and the private and public media in Zimbabwe. He also runs a climate change information blog: Climate Initiative Zimbabwe, where he posts features on climate change issues. Henri-Count has experience of working with children on environmental issues, having established Green Society Clubs in the Midlands and Bulawayo provinces in Zimbabwe. He partnered the Development Reality Institute in their school environmental clubs. Henri-Count has presented his work at academic and professional conferences.

Osée Kamga chairs the Programme d'Études journalistiques at the University of Sudbury, where he teaches journalism and public relations. He is also a research associate at the Audiovisual Media Lab for the study of Culture and Society (AMLAC&S) of the University of Ottawa. His field of interest covers social uses of new media, identity and culture, international communication, as well as ICTs for development. He is a specialist on Africa, and has written a number of papers on African development issues, including "The Weakness of 'Social Capital' as a Key Infringement to African Development: The Ivorian Example" in *Critical Arts: A Journal of South-North Cultural and Media Studies* no. 25 (3), September 2011. He organized and chaired the panel "African Future: Beyond Postcolonial Perspectives?" in the seventh biannual congress of the Association for Cultural Studies held in Paris in July 2012. Osée Kamga has been a guest speaker at the College of Arts and Sciences of the Case Western Reserve University in Cleveland, Ohio. He wrote *"Et si le développement nous trompait Le modèle ivoirien en point de mire"* (2006), an essay in which he articulates the issues of development policies for Africa, as well as the limits of the very concept of development.

Kosta Kostadinov is a consultant with the External Operational Communications Department of the World Bank. He specializes in development communication and social media and has designed, developed and managed the delivery of strategic communication courses for clients in more than two dozen countries in Asia, Africa and the Middle East. He is currently associated with the World Bank's Connect4Climate Initiative, which focuses primarily on global climate change communication. Prior to this, he worked as a communication specialist for multinational joint venture projects in the Middle East and as a broadcast journalist in Bulgaria. He was also an adjunct faculty member at Florida State University in Tallahassee, FL, and George Washington University in Washington, DC, where he taught courses in communication, film studies and legal issues of new media. He holds a master's degree in mass communication from Towson University, MD, and has completed coursework towards a PhD degree at Florida State University with concentration on development communication.

Patchanee Malikhao received her PhD in sociology from the University of Queensland in Australia, a master's of arts degree in mass communication from Thammasat University in Thailand, a master's of science degree in Printing Technology from Rochester Institute of Technology (RIT), Rochester, New York, and a bachelor of science (Hons.) in photographic science and printing technology from Chulalongkorn University in Thailand. She has worked and received extensive training in the fields of communication for social change, imaging technology, social science research methods and data analyses in Belgium, Australia and the United States. She has been a recipient of many scholarships and awards, including the Fulbright Scholarship, the Australian Postgraduate Award and the Outstanding Teacher Award. She worked from September 2008 to July 2011 as a researcher and a lecturer in the School of Public Health at the University of Massachusetts, Amherst, United States before she joined Thammasat University in Bangkok, Thailand, in July 2011. She has been Senior Fellow in the CSSC Center at UMass Amherst since 2010.

Maitreyee Mishra is Assistant Professor at Manipal Institute of Communication, Manipal University, India, and teaches environment and development communication, international communication and film studies. She has an MSc in communication for innovation and development from the University of Reading, United Kingdom. Maitreyee's research interests cover a broad range of topics, from environmental communication and philosophy, feminist studies, film theory and the intersection of film and literature. She has written and edited the publications of several NGOs and has presented papers on environmental communication and film studies at several international conferences. She recently presented a paper titled "Mass Media and the Construction of the Environment: A Discourse Analysis of Environmental News" in Orissa's newspapers and co-presented "Magical Realism and Sense of Reality" in Satoshi Kon's *Millennium Actress and Paprika*. In June 2012, her paper "Marriage, Devotion and Imprisonment: Women in Bimal Roy's *Devdas and Bandini*" was published in the Indian edition of the *Global Media Journal*. Having lived in many countries around the world, Maitreyee has had multi-cultural experiences which are reflected in her interests.

Kiran Prasad is Professor and Head, Department of Communication and Journalism, Sri Padmavati Mahila University, Tirupati, India. She was Commonwealth Visiting Research Fellow at the Centre for International Communication Research, Institute of Communication Studies, University of Leeds, UK, and Canadian Studies Research Fellow at the School of Journalism and Communication, Carleton University, Canada. She is the youngest ever recipient of the "State Best Teacher Award" from the Government of Andhra Pradesh, India, and recipient of several national awards for academic excellence. She has researched extensively in India, Bangladesh, Singapore, Malaysia, the Middle East, the United States, Canada and the United Kingdom on the interrelations between communication and development studies. She has published 20 books and over 100 research papers. Her recent books include *Media Law in India* (2011), *Communication for Development: Reinventing Theory and Action* (2009), *e-Journalism: New Media and News Media* (2009), *Media Law and Ethics: Readings in Communication Regulations* (2008) and *HIV and AIDS: Vulnerability of Women in Asia and Africa* (2008). She is Series Editor of *Empowering Women Worldwide*, New Delhi. She is also Country Director for India of the Asian Congress for Media & Communication (ACMC), Philippines, an international professional organization in the fields of media, communication and language education.

Boonlert Supadhiloke received a PhD in mass communication and an MA in journalism from the University of Wisconsin-Madison, United States, in 1976. He graduated with a BA in journalism (Honors) from Thammasat University, Thailand. He became the first Dean of the Faculty of Journalism and Mass Communication at Thammasat. He was transferred to the Government Public Relations Department, serving as Director of Radio Thailand, Deputy Director-General and Inspector-General. He was on deputation to serve the Food and Agriculture Organization of the United Nations as a development communication consultant during 1985–1988. He worked as Research Fellow at the East-West Communication Institute, Hawaii and was Minister-Counselor for Information, Royal Embassy of Thailand,

Washington DC, United States. He has published books and articles on a variety of research topics. He is currently an associate professor and a researcher at Bangkok University, Thailand.

Jagadish Thaker is a postdoctoral fellow at the Center for Culture-Centered Approach to Research and Evaluation (CARE), Department of Communications and New Media, National University of Singapore. His postdoctoral work examines the adaptation strategies of women farmers in South India. He is the co-author of "Climate Change in the Indian Mind", a report that assesses Indians' beliefs, attitudes, and policy support about climate change, and other sustainability-related issues. His primary research interests are in the fields of health communication, climate change communication, climate change and sustainable development discourses, media content analysis and strategic communication campaigns.

He holds a master's degree in English literature from the University of Hyderabad, India, and a PhD from George Mason University's Health and Strategic Communication Program for his research on the role of collective efficacy (perceptions of a group's collective abilities) in enhancing a community's adaptive capacity to climate change impacts.

Tran Van Dien is Rector of the Thai Nguyen University of Agriculture and Forestry in Thainguyen City, Vietnam.

Chin Saik Yoon has 35 years of wide-ranging, hands-on experience in information and communication action and research. During this time, he worked across developing and developed countries in Asia, the Americas and the Arab Region, at the community, national and regional policy levels. He is also an active participant in a number of international development networks and communities of practice related to participatory development communication. Chin was awarded the 2008 Communication for Social Change Award by the Centre for Communication and Social Change, University of Queensland, Australia, in recognition of his contributions to the field. He is currently the Publisher and Managing Director of Southbound, a scholarly publishing house specializing in books about communication and information for development and social change.

Acronyms

ABC	Abstinence, Be faithful and use a Condom
ABS	Access and Benefit Sharing
ACMAD	African Centre of Meteorological Applications for Development
ACTI	Agency of Telecommunications of Côte d'Ivoire
AEF	APEC Education Foundation
AIR	All India Radio
APCICT	Asia and Pacific Training Centre for Information and Communication Technologies for Development
APEC	Asia-Pacific Economic Cooperation
APEID	Asia-Pacific Programme of Educational Innovation for Development
BCC	Behavior Change Communication
BJP	Bharatiya Janata Party
BRAC	Bangladesh Rural Advancement Committee
CfD or C4D	Communication for Development
CDSC	Communication for Development and Social Change
COP	Conference of Parties
CSC	Communication for Social Change
CSD	Commission on Sustainable Development
CSO	Civil Society Organization
CSR	Corporate Social Responsibility
CSSC	Communication for Sustainable Social Change
CSSC&D	Communication for Sustainable Social Change and Development
DDS	Deccan Development Society
DSTV	Digital Satellite Television
DVD	Digital Versatile/Video Disc
ECA	Ethnographic Content Analysis
EFTA	European Fair Trade Association
EIA	Environmental Impact Assessment
FAO	Food and Agriculture Organization
FARR	Friends Association for Rural Reconstruction

FRA	Forest Rights Act
FGD	Focus Group Discussions
GAID	Global Alliance for ICT and Development
GDH	Gross Domestic Happiness
GDP	Gross Domestic Product
GED	Gender Development Index
GEM	Gender Equity Measure
GFATM	Global Fund to Fight AIDS, Tuberculosis and Malaria
GLTB	Gay, Lesbian, Transgender and Bisexual
GNH	Gross National Happiness
GNP	Gross National Product
HDI	Human Development Index
HIV/AIDS	Human Immunodeficiency Virus/Acquired Immunodeficiency Syndrome
HPII	Happy Planet Index
HPI	Human Poverty Index
ICTs	Information and Communication Technologies
ICTD	Information and Communication Technologies for Development
IDCO	Industrial Infrastructure Development Corporation of Orissa
IMF	International Monetary Fund
INEXSK	Infrastructure, Experience, Skills, Knowledge
IT	Information Technology
ITU	International Telecommunications Union
IUCN	International Union for Conservation of Nature
KAP	Knowledge, Attitude and Behavior
LDCs	Least Developed Countries
MDGs	Millennium Development Goals
MKSS	Mazdoor Kisan Shakti Sangathan
MoUs	Memoranda of Understandings
M&E	Monitoring and Evaluation
NAPM	National Alliance of People's Movements
NBA	Narmada Bachao Andolan
NCPRI	National Campaign for People's Right to Information
NEWS	Network of European Worldshops
NESAC	National Economic and Social Advisory Council
NGO	Non-Governmental Organization
NICTs	New Information and Communication Technologies
NMA	Northern Mountainous Area
NWICO	New World Information and Communication Order

PBS	Public Broadcasting Service
PEPFAR	United States of America President's Emergency Plan for AIDS Relief
PDA	Population & Community Development Association
PPP	Public-Private Partnership
PPSS	Posco Pratirodh Sangram Samiti
PSA	Public Service Advertisement
PTCPs	Post, Telecom and Cultural Points
R&D	Research and Development
RME	Research, Monitoring and Evaluation
RTC	Right to Communicate
RURANET/ICD	Rural Radio Network/Information Centers for Development
SCOT	Social Construction of Technology
SD	Sustainable Development
SE	Sufficiency Economy
SEMRRU	Self Managed Solar Rural Radio Unit
SEZ	Special Economic Zone
SDG	Sustainable Development Goals
SHG	Self Help Group
SME	Small and Medium-sized Enterprise
STI	Sexually Transmitted Infections
STD	Sexually Transmitted Diseases
TANUVAS	TN University of Veterinary and Animal Sciences
TB	Tuberculosis
TINA	There is no Alternative
TUAF	Thai Nguyen University of Agriculture and Forestry (Vietnam)
UDHR	Universal Declaration of Human Rights
UN	United Nations
UNAIDS	Joint United Nations Programme on HIV/AIDS
UNCDP	United Nations Capital Development Programme
UNCED	United Nations Conference on Environment and Development
UNCSD	United Nations Conference on Sustainable Development
UNCSTD	United Nations Commission on Science and Technology for Development
UNCTAD	United Nations Conference on Trade and Development

UNDP	United Nations Development Programme
UNEP	United Nations Environment Programme
UNESCO	United Nations Educational, Scientific and Cultural Organization
UNFCCC	United Nations Framework Convention on Climate Change
UNFPA	United Nations Population Fund
UNICEF	United Nations Children's Fund
UNIFEM	United Nations Development Fund for Women
USAID	United States Agency for International Development
VICs	Village Information Centres
VVJM	Visthapan Virodhi Jan Manch
WB	World Bank
WCC	World Council of Churches
WCED	World Commission on Environment and Development
WFTO	World Fair Trade Organization
WHO	World Health Organization
WSIS	World Summit on the Information Society
WMO	World Meteorological Organization
WTO	World Trade Organization
WWF	World Wildlife Fund

Introduction: Imperatives for a Sustainable Future

Jan Servaes

The problems we face today regarding climate change, terrorism, pandemics and deep fractures in world trade, commerce and politics are unlikely to be solved quickly. Let's list some of these "problems" or "risks":

- Infectious diseases such as HIV, AIDS, severe acute respiratory syndrome (SARS), Avian flu and Ebola, for which cures have not been developed and where the main instrument of control in the event of a pandemic is timely and effective crisis communication.
- Climate change may lead to the permanent flooding of low-lying regions.
- Disruption of potable water supply to millions around the world, as the glaciers in the mountains shrink and deplete the water reserves of major population centres.
- Transformation of parts of the global economy to services that trade valuable information secured via intellectual property regimes.
- Political tensions and conflict caused by confrontation and competition among religious and cultural groups across the world.
- Armed conflict and genocide.
- Disruption in affordable, locally produced food supplies causing widespread food insecurity.
- Decline in oil and gas supplies as hydrocarbon reserves peak and the price of energy increases exponentially.
- Disruption in international finance and trade causing the world economy to crash, leading to a prolonged depression.
- Erosion of human rights as the security of states and communities are being threatened by political, economic, financial and ecological challenges.
- Erosion and displacement of deeply rooted cultural and religious beliefs and values of communities.

Whereas in the past we were able to increase food production over a few cropping cycles, or establish income-generating ventures in a couple of years, the "new" problems we face may take years and, in the case of climate change, several generations for the world community to resolve. How do we build consensus and muster the altruistic intent of the present generation to consume less, de-escalate conflict and subject ourselves to medical research so that future generations who will exist long after we are gone may inherit a habitable planet?

The tried and tested methods of agriculture extension, social mobilization, community participation and multi-lateral negotiation are unlikely to succeed on their own as these systemic problems grow in their severity and people submit to innate human instincts for self-preservation and compete even more keenly for rapidly dwindling natural resources, ratchet up violence, resist Hippocratic principles to share limited supplies of vaccines and medicines, hoard energy and water, and close markets to international commerce.

We do not have appropriate strategies to begin addressing these "new" and highly complex challenges.

Sustainable development: Definitions and perspectives[1]

In the last 20 years, sustainable development (SD) has emerged as one of the most prominent development paradigms. In 1987, the World Commission on Environment and Development (WCED) – in short: Brundtland Commission – concluded that "sustainable development is development that meets the needs of the present without compromising the ability of future generations to meet their own needs." SD is seen as a means of enhancing decision-making so that it provides a more comprehensive assessment of the many multi-dimensional problems society faces (Ekins, 1993; Elliott, 1994; Lele, 1991; Taylor, 1996).

What is required is an evaluation framework for categorizing programs, projects, policies, and/or decisions as having sustainability potential (Lennie & Tacchi, 2010, 2013). Much of the current sustainability debate in the West remains based on the acceptance of the existing capitalist system as it is.

> As a result, solutions to the multitude of problems we face, from climate change and biodiversity loss to resource and water scarcity, ten to be built around iterative change. While those modest moves forward are welcome and create a foundation for taking the next step, many of us recognize the risk that we are fiddling while Rome burns.
>
> (Confino, 2011; see also WEF 2012)

In other words, rather than looking at these multi-dimensional and complex problems in isolation and trying to "solve" them in short-term, piecemeal, politically negotiated baby steps, some "bold" thinking and action is required. We need creative and critical thinking and action that value the deeper meaning of sustainability, as the ability to connect social, environmental, ethical, cultural and economic issues in a holistic and participatory way.

Three dimensions are generally recognized as the "pillars" of SD: economic, environmental and social.

> The essence of sustainability therefore, is to take the contextual features of economy, society, and environment – the uncertainty, the multiple competing values, and the distrust among various interest groups – as givens and go on to design a process that guides concerned groups to seek out and ask the right questions as a preventative approach to environmentally and socially regrettable undertakings.
>
> (Flint, 2007: IV)

Many authors (see, for instance, Amin, 2006; Bhambra, 2007; Blewitt, 2008; de Sousa, 2007; Esteva & Prakah, 1998; Gawor, 2008; Sanyawiwat, 2003; Shah, 2005) doubt whether the "Western" perspective, as represented by the Brundtland Commission, has fully embraced a vision of SD that looks far beyond slogans of corporate responsibility or transparency. They urge us to look for guidance elsewhere as well. Therefore, we wish to briefly complement the Western perspective with an "Eastern" or Buddhist perspective, as presented by the Thai philosophers and social critics Sulak Sivaraksa and Phra Dhammapidhok.

A "Western" perspective: the Brundtland Commission

An interest in SD gained momentum at the convening of the WCED at the United Nations in 1983 to address growing concern "with the problems of protecting and enhancing the environment". The 1987 report by the Brundtland Commission, *Our Common Future*, was one of the first cohesive reports to consider economic and social development in terms of sustainability. Core issues and necessary conditions for SD, as identified by the WCED, are: population and development; food security; species and ecosystems; energy; industry; and the urban challenge.

In 1992, *Agenda 21*, a plan of action to produce international and national SD strategies, was adopted by more than 178 governments at the United Nations Conference on Environment and Development (UNCED) held in Rio de Janeiro, Brazil, 3 to 14 June 1992. This led to the creation of the Commission on Sustainable Development (CSD) later that year to ensure effective follow-up of UNCED. In 2002, the United Nations World Summit on Sustainable Development was convened to assess the effectiveness of Agenda 21. The five areas discussed at this conference were:

1. water and sanitation;
2. energy;
3. human health;
4. agricultural productivity; and
5. biodiversity and ecosystem management.

Pursuit of this kind of SD requires:

- a political system that secures effective citizen participation in decision-making;
- an economic system that provides for solutions for the tensions arising from disharmonious development;
- a production system that respects the obligation to preserve the ecological base for development;
- a technological system that fosters sustainable patterns of trade and finance;
- an administrative system that is flexible and has the capacity for self-correction; and
- a communication system that gets this organized and accepted by all parties concerned at all levels of society.

It is unclear, however, what has gone beyond ratified agreements and stated commitments. Shah (2005) notes that since the commitments made in 1992, little has changed in terms of global poverty. The rising popularity of the term through conferences, protocols and agreements has ironically blurred the definition of sustainability and made the understanding of it vague (Hull, 2008). Hull suggests that a Western initiate model of development has emphasized economic growth through industrialization and technological growth. A huge deterrent to sustainability is global turbo capitalism, where "society serves the economy and not vice versa" (Hull, 2008: 74). Gawor (2008) suggests that SD should be understood as an alternative to "development megatrends of the present, including globalization processes denoting the need to change the previous values, which contributed to the ride of Euro-American industrial-technological civilization" (Gawor, 2008: 131). Gawor suggests that anti-globalization movements, including activism against the World Trade Organization and International Monetary Fund are a cry for a new alternative (see also Held & McGrew, 2007).

Kosta Kostadinov and Jagadish Thaker, in their chapter in this volume, summarize what has happened over the past decade, culminating in the so-called Rio+20 Summit of June 2012.

An "Eastern" Buddhist perspective

Phra Dhammapidhok (Payutto, 1998), a famous Buddhist monk and philosopher, points out that SD from a Western perspective lacks the human development dimension. He states that Western ideology emphasizes "competition". Therefore, the concept of "compromising" is used in the above WCED definition. To compromise is to make a deal between different parties, where each party gives up part of their demands and, therefore, lessens the needs of all parties. If the other parties do not want to compromise, you have to compromise your own needs and that will lead to frustration. Development won't be sustained if people are not happy.

He consequently reaches the conclusion that the Western perception of, and road to, sustainability, based on Western ethics, leads development into a *cul-de-sac*.

From a Buddhist perspective, sustainability concerns ecology, economy and *evolvability*. The concept of "evolvability" means the potential of human beings to develop themselves into less selfish persons. The main core of SD is to encourage and convince human beings to live in harmony with their environment, not to control or destroy it. If humans have been socialized correctly, they will express the correct attitude toward nature and the environment and act accordingly. He argues that:

> A correct relation system of developed mankind is the acceptance of the fact that human-being is part of the existence of nature and relates to its ecology. Human-being should develop itself to have a higher capacity to help his fellows and other species in the natural domain; to live in a harmonious way and lessen exploitations in order to contribute to a happier world.
>
> (Payutto, 1998: 189)

This holistic approach relates to cultural development in three dimensions:

behaviours and lifestyles which do not harm nature;
minds in line with (Eastern) ethics, stability of mind, motivation to see other creatures as companions; and
wisdom, which includes knowledge and understanding, attitude, norm and values in order to live in harmony with nature.

Different perspectives (such as the TERMS [Technological, Economic, natural Resource, Mental and Sociocultural] approach developed in Thailand, which builds on Buddhist principles and the "efficiency economy" concept outlined by King Bhumibol – see Supadhiloke, 2010; Servaes & Malikhao, 2007a,b; and below) have, over the years, influenced this holistic and integrated vision of SD. Khampa (2009), Supadhiloke (2010) and Sivaraksa (2010) also explore the Bhutanese Gross National Happiness Index as a viable way to SD and a realistic alternative to the Western concept. Sivaraksa (2010: 66) lists the following indicators of happiness:

- the degree of trust, social capital, cultural continuity and social solidarity;
- the general level of spiritual development and emotional intelligence;
- the degree to which basic needs are satisfied;
- access to and the ability to benefit from health care and education; and
- the level of environmental integrity, including species loss or gain, pollution and environmental degradation.

Sivaraksa argues that these indicators need to be further operationalized. A task which Khampa (2009) is currently involved in on behalf of the Bhutanese government. The key is "to create indicators that become instruments of liberation" (Sivaraksa, 2010: 67).

A "middle way"?

It may be relevant to emphasize that the above "Eastern" perspective is not "uniquely" Eastern or Asian, as it has been promoted in other parts of the world as well. For instance, in the late 1970s, the Dag

Hammerskjold Foundation advocated three foundations for "another" development or SD:

- Another Development is geared to the satisfaction of needs, beginning with the eradication of poverty;
- Another Development is endogenous and self-reliant; and
- Another Development is in harmony with the physical and cultural ecology (Nerfin, 1977).

An interesting, often overlooked, contribution has been made by Manfred Max-Neef (1991), who, as the Executive Director of the Chilean Development Alternatives Centre, and with the support of the Dag Hammerskjold Foundation, edited a selection of chapters which together resulted in a transdisciplinary model for "human scale development" with self-reliance among

- human beings, nature and technology;
- the personal and the social;
- the micro and the macro;
- planning and autonomy; and
- the state and civil society

as central to empower groups and social actors: "The fundamental issue is to enable people from their many small and heterogeneous spaces to set up, sustain and develop their own projects" (Max-Neef, 1991: 85).

More recently, the World Commission on Culture and Development, chaired by Javier Pérez de Cuéllar (1995), started from similar assumptions. It argued that development divorced from its human or cultural context is growth without a soul. This means that culture cannot ultimately be reduced to a subsidiary position as a mere promoter of economic growth. The report goes on to argue that "governments cannot determine a people's culture: indeed, they are partly determined by it" (De Cuéllar, 1995: 15).

The basic principle should be

> the fostering of respect for all cultures whose values are tolerant of others. Respect goes beyond tolerance and implies a positive attitude to other people and a rejoicing in their culture. Social peace is necessary for human development: in turn it requires that differences between cultures be regarded not as something alien and

> unacceptable or hateful, but as experiments in ways of living together that contain valuable lessons and information for all.
>
> (De Cuéllar, 1995: 25)

The Human Development Report 2004 and the *United Nations Millennium Declaration* (2000) advocate these principles of cultural liberty and cultural respect in today's diverse world for similar reasons: "The central issue in cultural liberty is the capability of people to live as they would choose, with adequate opportunity to consider other options" (UNDP, 2004: 17). The United Nations Millennium Declaration (2000) promotes the following principles and values:

- *Freedom*: Men and women have the right to live their lives and raise their children in dignity, free from hunger and from the fear of violence, oppression or injustice. Democratic and participatory governance based on the will of the people best assures these rights.
- *Equality*: No individual and no nation must be denied the opportunity to benefit from development. The equal rights and opportunities of women and men must be assured.
- *Solidarity*: Global challenges must be managed in a way that distributes the costs and burdens fairly in accordance with the basic principles of equity and social justice. Those who suffer or who benefit least deserve help from those who benefit most.
- *Tolerance*: Human beings must respect one another, in all their diversity of belief, culture and language. Differences within and between societies should be neither feared nor repressed, but cherished as a precious asset of humanity. A culture of peace and dialogue among all civilizations should be actively promoted.
- *Respect for nature*: Prudence must be shown in the management of all living species and natural resources, in accordance with the precepts of SD. Only in this way can the immeasurable riches provided to us by nature be preserved and passed on to our descendants. The current unsustainable patterns of production and consumption must be changed in the interest of our future welfare and that of our descendants.
- *Shared responsibility*: Responsibility for managing worldwide economic and social development, as well as threats to international peace and security, must be shared among the nations of the world and should be exercised multilaterally. As the most universal and most representative organization in the world, the United Nations must play the central role.

Therefore, in contrast to the more economically and politically oriented approach in the traditional perspectives on SD, the central idea in alternative, more culturally oriented versions is that there is *no universal development model* which leads to sustainability at all levels of society and the world, that development is an integral, multi-dimensional and dialectic process that can differ from society to society, community to community, context to context. In other words, each society and community must attempt to delineate its own strategy for SD (Servaes, 1999). This implies that the development problem is relative and that no one society can contend that it is "developed" in every respect.

The so-called Copenhagen Consensus project[2] is worth mentioning in this context. Though still dominated by economic perspectives and researchers (some of them Nobel Prize winners), the panel of experts evaluated a large number of development recommendations, drawn from assessments by United Nations agencies, and identified ten core challenges for the future:

- Civil conflicts
- Climate change
- Communicable diseases
- Education
- Financial stability
- Governance
- Hunger and malnutrition
- Migration
- Trade reform
- Water and sanitation

The major challenge identified was the fight against HIV/AIDS (human immunodeficiency virus/acquired immunodeficiency syndrome). See Patchanee Malikhao's contribution for more details.

Therefore, we believe that the scope and degree of interdependency must be studied in relation to the content of the concept of development. Where previous perspectives did not succeed in reconciling economic growth with social justice, an attempt should be made to approach problems of freedom and justice from the relationship of tension between the individual and society, and limits of growth and sustainability are seen as inherent to the interaction between society and its physical and cultural ecology.

The multi-dimensionality of sustainable development

The concept of SD has been further addressed from at least three dimensions:

1. as a process;
2. at different levels; and
3. with different contents.

Open, inclusive and participatory communication and information processes are fundamental for successful SD (Wilson, 2007). Furthermore, "when communities articulate their own agendas, they are more likely to achieve positive changes in attitudes, behaviors, and access to opportunities" (Reardon, 2003). Wilson offers four key elements that will promote SD: "Equitable and inclusive political processes, national and international governance processes that are effective, responsive, and accountable, supporting engaged citizens and dynamic civil society, and generating inclusive economic growth, sustainable livelihoods and transparent, efficient markets" (Wilson, 2007).

Chen (2001) and Tremblay (2007) indicate that the goal of SD is to pursue "regional balanced-development", suggesting that a large challenge is to strike harmony between the environment and the expansion of science and technology. On the one hand, protection of resources is key; however, fair global distribution of resources is contradictory to the structure of competition, which encourages survival of the fittest, with a privileged few gaining access to resources. The goal of sustainability should not be to substitute man-made or artificial capital by natural resources but to have each complement the other. This is what is known as strong sustainability (Horbach, 2005).

Skowrownski (2008: 119) calls for "environmentally friendly socio-economic development that takes account of the finite nature of environmental resources and possibilities." He distinguishes between culture at its essence, and material forms of culture, suggesting that civilizational culture need not be based on mastering the natural environment and shaping nature. The two basic approaches to SD are, first, approaching a balance or reconciliation of traditional economic growth with ecological and environmental conditionings and, second, a philosophy or ideology that conceptualizes civilization in a holistic manner.

Mannberg & Wihlborg (2008) acknowledge that global and local visions of sustainability are often unaligned. They suggest that the root of SD is in fact in local, well-functioning planning processes that are decentralized, and grassroots. They use the concept of communicative planning to suggest that a socially sustainable society is one where participation is part of planning processes. "Communicative planning is characterized by a view of planning as a long-term process in which the focus lies not only on the planning object, but also on the process as such and on communication rather than calculations" (Mannberg & Wihlborg, 2008: 36). It is participation that allows sustainability at the local level, where locals are part of the process of defining what is sustainable for them.

In sum, over the years different perspectives – based on both "Western" and "Eastern" philosophical starting points – have resulted in a more holistic and integrated vision of SD. Each society and community must attempt to delineate its own strategy to SD, starting with the resources and "capitals" available (not only physical, financial and environmental but also human, social and institutional capitals) and considering the needs and views of the people concerned.

SD implies a participatory, multi-stakeholder approach to policy making and implementation, mobilizing public and private resources for development and making use of the knowledge, skills and energy of all social groups concerned with the future of the planet and its people. Within this framework, communication and information play a strategic and fundamental role by

- contributing to the interplay of different development factors;
- improving the sharing of knowledge and information; and
- encouraging the participation of all concerned.

Sustainable development at the community level

In view of the above expressed need to start at a local or community level, we will present the Thai double-tiered so-called TERMS model of rural community self-reliance (Sanyawiwat, 2003) as an interesting integrated framework. TERMS stands for Technological, Economic, natural Resource, Mental and Sociocultural. This model is the result of extensive research, which the Thai National Research Council commissioned the Science and Technology Institute to conduct. More than 50 academics were involved, who came from governmental bureaus, universities and the private sector. Community leaders from five villages were also involved. It took them more than seven years to arrive at what is now called the *Thai concept of community development*.

This model views self-reliance of a community as a goal of community development. Self-reliance of a community can be established if the following dimensions are taken into account:

- TERMS factors;
- a development and self-reliant process based on balance, ability and networking (BAN), with these three factors running in parallel with the TERMS factors and community management, all in balance;
- a participatory action research approach in which the facilitators and villagers collaborate through discussion, planning, evaluation or research at all times;
- a resocialization and conscientization process (*Khit pen* in Thai), which makes the people value Thainess, the Thai identity as well as Thai culture and folk wisdom so as to promote the Thai way of living.

The functional matrix of these factors is made visible in Table I.1.

Overview of development and communication paradigms

To fully understand where and how the issue of sustainability has gradually come into the picture, a brief overview of the past and present of the development and communication discourses may be fitting.

Development paradigms

1. After the Second World War, the founding of the United Nations stimulated relations among sovereign states, especially the North Atlantic Nations and the so-called developing nations, including the new states emerging out of a colonial past. During the Cold War period, the superpowers – the United States and the former Soviet Union – tried to expand their own interests to the so-called Third World or developing countries. In fact, the United States was defining development and social change as the replica of its own political-economic system and opening the way for the transnational corporations. At the same time, the developing countries saw the "welfare state" of the North Atlantic Nations as the ultimate goal of development. These nations were attracted by the new technology transfer and the model of a centralized state with careful economic planning and centrally directed development bureaucracies for agriculture, education and health as the most effective strategies to catch up with those industrialized countries (McMichael, 2008; Nederveen Pieterse, 2010).

Table I.1 The TERMS model of community self-reliance

Self-reliance	Technological	Economic	Resource	Mental	Socio-cultural
Technology	Use technologies that are: appropriate for the rural environment modern and controllable researched and developed indigenously	Establish: a production base factors of production competitiveness a strong development base	Add value to resources Protect the environment Restore resources	Build scientific consciousness Improve human resources in science and technology	Build a social development base Maintain balance of social change factors and social stability Preserve social stability and solidarity
Economy	Support: choice of technology technological progress technological self-reliance distribution of technology	Create an equilibrium state Promote continuous development Develop ability to compete externally Develop ability to save and invest Cooperate in production and marketing activities	Use local resources Increase value added to resources Use resources efficiently and in balance Restore and recycle resources	Create consciousness about: trading quality of life social equity saving and appropriate investment	Create jobs and incomes Build solidarity and social balance Create a stable and fair economic system Integrate economic profits into society
Resources	Develop: a technological development base multiplicity in technology use technological innovations	Develop: a production base and appropriate entrepreneurship factors of production a sustainable development base	Maintain ecological balance in resource use Develop recyclable resources	Create ecological consciousness Create consciousness of sustainable economic development Support quality of life	Build social groups Preserve and pass on traditions and culture Develop discipline in resource allocation Build social solidarity

Table I.1 (Continued)

Self-reliance	Technological	Economic	Resource	Mental	Socio-cultural
Mind	Develop capability to use technology Control and monitor technology use	Build labour factor of production Develop entrepreneurship Develop consumer knowledge about economizing Seize economic opportunities Develop creators of economic organizations	Develop: knowledge of resource use knowledge of resource conservation capability to restore resources a love for nature	Develop a self-reliance attitude Develop capacity for self-development Be knowledgeable and able to apply knowledge Nurture good qualities Abstain from intoxicants and gambling Be diligent	Build consciousness of social membership Participate in community activities Motivate self to progress Develop creators of social organizations Build social cohesion
Society	Assess need for technology Support and develop appropriate technology Determine type and form of technology needed	Create demand for products and services Create social organizations Create values and norms of economic self-reliance Administrate production and marketing activities	Share and conserve resources Restore resources Care for the environment and the community	Establish order and discipline Respect discipline and social rules Create consciousness of social cohesion Cultivate a belief in creating social benefits	Develop leadership qualities Foster social solidarity Create social organizations Build knowledge and current information

This mainly economic-oriented view, characterized by endogenism and evolutionism, ultimately resulted in modernization and growth theory. It sees development as an unilinear, evolutionary process and defines the state of underdevelopment in terms of observable quantitative differences between so-called poor and rich countries on the one hand, and traditional and modern societies on the other hand (for more details, see Servaes, 1999, 2003, 2008, 2012).

2. As a result of the general intellectual "revolution" that took place in the mid-1960s, this Euro- or ethnocentric perspective on development was challenged by Latin American social scientists, and a theory dealing with dependency and underdevelopment was born (see one of the "classics": Cardoso & Faletto, 1969). The dependency approach formed part of a general structuralist re-orientation in the social sciences. The "dependistas" were primarily concerned with the effects of dependency in peripheral countries, but implicit in their analysis was the idea that development and underdevelopment must be understood in the context of a world system (Chew & Denemark, 1996).

   This dependency paradigm played an important role in the movement for a New International Economic Order (NIEO) and New World Information and Communication Order (NWICO) from the late 1960s to the early 1980s. At that time, the new states in Africa, Asia and the success of socialist and popular movements in Cuba, China, Chile and other countries provided the goals for political, economic and cultural self-determination within the international community of nations. These new nations shared the ideas of being independent from the superpowers and moved to form the Non-Aligned Nations. The Non-Aligned Movement defined development as political struggle.
3. Since the demarcation of the First, Second and Third Worlds has broken down and the cross-over center-periphery can be found in every region, there was a need for a new concept of development which emphasized cultural identity and multi-dimensionality (further discussed in Barbero, 1993; Canclini, 1993, 1995; De Cuellar, 1995; Robertson, 1992; Sen, 2004; Servaes, 1999, 2008; Tomlinson, 1999).

From the criticism of the two paradigms above, particularly that of the dependency approach, a new viewpoint on development and social change came to the forefront. The common starting point here is the examination of the changes from "bottom-up", from the self-development of the local community. The basic assumption was that there are no countries or communities that function completely autonomously and that are completely self-sufficient, nor are there any nations whose development is exclusively determined by external factors. Every society is dependent in one way or another, both in form and in degree. Thus, a framework was sought within which both the center and the periphery could be studied separately and in their mutual relationship, both at global, national and local levels.

More attention was also being paid to the content of development, which implied a more normative, holistic and ecological approach. "Another development" questions whether "developed" countries are in fact developed and whether this genre of progress is sustainable or desirable. It favors a multiplicity of approaches based on the context and the basic, felt needs, and the empowerment of the most oppressed sectors of various societies at divergent levels. A main thesis is that change must be structural and occur at multiple levels in order to achieve sustainable ends.

Communication paradigms

1. The above more general typology of the so-called development paradigms can also be found at the communication and culture level. The communication media are, in the context of development, generally used to support development initiatives by the dissemination of messages that encourage the public to support development-oriented projects. Although development strategies in developing countries diverge widely, the usual pattern for broadcasting and the press has been predominantly the same: informing the population about projects, illustrating the advantages of these projects, and recommending that they be supported. A typical example of such a strategy is situated in the area of family planning, where communications such as posters, pamphlets, radio and television attempt to persuade the public to accept birth-control methods. Similar strategies are used on campaigns regarding health and nutrition, agricultural projects, education, HIV/AIDS prevention and so on.

This model sees the communication process mainly as a message going from a sender to a receiver. This hierarchic view on communication can be summarized in Laswell's classic formula – "Who says What through Which channel to Whom with What effect?" – and dates back to (mainly American) research on (political) campaigns and diffusions in the late 1940s and 1950s (Lerner, 1958; Lerner & Schramm, 1967; Schramm, 1964; Schramm & Lerner, 1976). The most important hypothesis of Lerner's study was that a high level of empathy is "the predominant personal style only in modern society, which is distinctively industrial, urban, literate and participant" (Lerner, 1958: 50). Empathic persons have a higher degree of mobility, meaning a high capacity for change, and were more future oriented and rational than so-called traditional people. According to Lerner, the general psychological conditions captured by the concept of empathy stimulated mobility and urbanization, which, in turn, increases literacy and consequently economic and political participation – all essential to the modernization process. The media would serve to stimulate, in direct and indirect ways, the conditions of "psychological mobility" that are so crucial to economic development.

Schramm stressed the importance of developing modern communication infrastructures and, based on the weakness of the private sector and the lack of economic resources in most developing countries, he advocated that the state, or governments, should play a leading role in infrastructural development. While this obviously enhanced the role of the state vis-à-vis the media, since the government under Schramm's proposals would own and control the communications media, other theorists suggested the training of media professionals and journalists in the traditions of Western media theory. As such, the over-riding interest was in gaining access to the technology as quickly as possible, even if this meant turning a blind eye to the power relations that would shape the transfer and deployment of technology, as in the case of Schramm's proposal for government development of national media systems.

Referring to the agricultural extension work done in the United States in the first half of the past century and borrowing concepts from rural sociology, the American scholar Rogers (1962, 1976, 1986, 2003) is said to be the person who introduced the diffusion theory in the context of development. Modernization is here conceived as a process of diffusion whereby individuals move from a traditional way of life to a different, more technically developed and more rapidly changing way of life. Building primarily on sociological research in agrarian societies, Rogers stressed the adoption and diffusion processes of cultural innovation. This approach is therefore concerned with the process of diffusion and adoption of innovations in a systematic and planned way. Mass media are important in spreading awareness of new possibilities and practices, but at the stage where decisions are being made about whether to adopt or not to adopt them, personal communication is far more likely to be influential. Therefore, the general conclusion of this line of thought is that mass communication is less likely than personal influence to have a direct effect on social behavior.

Volker Hoffmann's (2007) close reading of the five editions of *Diffusion of Innovations* is a must-read for all of us. Hoffmann argues that Rogers' theory led to more and more inconsistencies and internal contradictions from edition to edition.

2. Newer perspectives on development communication claim that the diffusion and modernization perspective is a limited view of development communication. Juan Diaz Bordenave (1977) and Luis Ramiro Beltran (1976) were among the first Latin American scholars who fundamentally questioned the capacity of the diffusionist communication model to modernize Latin American societies.

They argued that this diffusion model is a vertical or one-way perspective on communication, and that development will accelerate mainly through active involvement in the process of the communication itself. Research has shown that, while groups of the public can obtain information from impersonal sources such as radio, television and, nowadays, the Internet, this information has relatively little effect on behavioral changes. And development envisions precisely such change. Similar research has led to the conclusion that more is learned from interpersonal contacts and from mass communication techniques that are based on them. On the lowest level, before people can discuss and resolve problems, they must be informed of the facts, information that the media provide nationally as well as regionally and locally. At the same time, the public, if the media are sufficiently accessible, can make its information needs known.

Communication theories such as the "diffusion of innovations", the "two-step-flow", or the "extension" approaches are quite congruent with the above modernization theory. The elitist, vertical or top-down orientation of the diffusion model is obvious (for more details, see Fraser & Restrepo-Estrada, 1998; Mody, 1997, 2003; Servaes 2003, 2008).

One of the more useful attempts to theorize the means by which cultural/media dependency works through the modern forms of electronic communication is that offered by Oliver Boyd-Barrett (1977). According to Boyd-Barrett, the international communication process consists of four major interrelated components:

1. the shape of the communication vehicle, involving a specific technology for the distribution and consumption of media products;
2. a set of industrial arrangements that organize the relations between media finance, production, distribution and consumption;
3. a body of values about ideal practice; and
4. specific media contents.

Herb Schiller (1976), one of the most prolific American communication scholars on dependency issues, called it "cultural imperialism". He studied the links between the United States military-industrial complex and the media industry and its consequences for culture and communication worldwide. Schiller's definition of cultural imperialism is broad: "The sum of the processes by which a society is brought into the modern world system and how its dominating stratum is attracted, pressured, forced, and sometimes bribed into shaping social institutions to correspond to, or even promote, the values and structures of the dominating centre of the system" (Schiller, 1976: 9).

3. The participatory model, on the other hand, incorporates the concepts in the framework of multiplicity. It stresses the importance of cultural identity of local communities and of democratization and participation at all levels – global, international, national, local and individual. It points to a strategy, not merely inclusive of, but largely emanating from, the traditional "receivers". Paulo Freire (1983:76) refers to this as the right of all people to individually and collectively speak their word: "This is not the privilege of some few men, but the right of every (wo)man. Consequently, no one can say a true word alone – nor can he say it for another, in a prescriptive act which robs others of their words."

In order to share information, knowledge, trust, commitment and the right attitude in development projects, participation is very important in any decision-making process for development. Therefore, the International Commission for the Study of Communication Problems, chaired by the late Sean MacBride, argued that "this calls for a new attitude for overcoming stereotyped thinking and to promote more understanding of diversity and plurality, with full respect for the dignity and equality of peoples living in different conditions and acting in different ways" (MacBride, 1980:254). This model stresses reciprocal collaboration throughout all levels of participation (for more details, see Jacobson & Servaes, 1999; Papa, Singhal & Papa, 2006; Servaes, Jacobson & White, 1996).

In line with this vision, at the end of the 1980s the participatory approach became a key feature in the applications of Communication for Sustainable Social Change and Development (CSSC&D). CSSC&D is about dialogue, participation and the sharing of knowledge and information. It takes into account the needs and capacities of all concerned through the integrated and participatory use of communication processes, media and channels. It works by:

- *facilitating participation*: giving a voice to different stakeholders to engage in the decision-making process;
- *making information understandable and meaningful*. It includes explaining and conveying information for the purposes of training, exchange of experience and sharing of know-how and technology; and
- *fostering policy acceptance*: enacting and promoting policies that increase rural people's access to services and resources.

Also, these newer approaches argue, the point of departure must be the *community* (see, for instance, Fuglesang, 1982; Geertz, 1973; Omoto, 2005; Servaes & Liu, 2007). It is at the community level that the problems of living conditions are discussed, and interactions with other communities are elicited. The most developed form of participation is self-management. This principle implies the right to participate in the planning and production of media content. However, not everyone wants to or must be involved in its practical implementation. More important is that participation is made possible in the decision-making regarding the subjects treated in the messages and regarding the selection procedures.

One of the fundamental hindrances to the decision to adopt the participation strategy is that it threatens existing hierarchies. However, participation does not imply that there is no longer a role for development specialists, planners, and institutional leaders. It only means that the viewpoint of local groups of the public is considered before the resources for development projects are allocated and distributed, and that suggestions for changes in the policy are taken into consideration.

Research priorities

This rather theoretical overview of development and communication paradigms needs to be complemented with an assessment of the research priorities in the CSSC&D field over the past decades. Three scholars who have been assessing the "richness" of the CSSC&D field in an historical perspective are Jo Ellen Fair, Hemant Shah and Christine Ogan. They each – either alone or with their graduate students, by examining journal articles, books and book chapters – wanted to highlight the directions which the research on CSSC took in different periods; from 1958 to 1986, from 1987 to 1996, from 1997 to 2005 and from 1998 to 2007.

In the 1958–1986 period, models predicting either powerful or limited effects informed the research.

> Communication has been a key element in the West's project of developing the Third World. In the one-and-a-half decades after Lerner's influential 1958 study of communication and development in the Middle East, communication researchers assumed that the introduction of media and certain types of educational, political, and economic information into a social system could transform individuals and societies from traditional to modern. Conceived as having fairly direct and powerful effects on Third World audiences, the media were seen as magic multipliers, able to accelerate and magnify the benefits of development.
>
> (Fair, 1989: 145)

> In the 1987–1996 period, Lerner's modernization model completely disappears. Instead, the most frequently used theoretical framework is participatory development, an optimist postmodern orientation, which is almost the polar opposite of Lerner who viewed mass communication as playing a top-down role in social change. Also vanishing from research in this latter period is the two-step flow model, which was drawn upon by modernization scholars.
>
> (Fair & Shah, 1997:10)

The two more recent periods, 1997–2005 and 1998–2007, which partly overlap, provide new findings which may be surprising to some.

I quote from Shah's essay first:

> First, Lerner's model of media and development has reappeared in the 1997–2005 time period after totally disappearing in the 1987–1996 period. Second, only two other theories from the traditional US-based behavioral science approach, social learning theory and knowledge gap, appear in the 1997–2005 period... The third trend to note is

> that the two most prominently mentioned theories in 1997–2005 – participatory communication and social learning – reflect two popular development communication project orientations that were mentioned as innovations in the 1987–1996 study: participatory development and edu-tainment.
>
> (Shah, 2007: 13)

Shah explains the persistence of "old" ideas, especially Lerner's model (1958, 1977), from a technological deterministic perspective: "Each new technological innovation in the postcolonial world since 1958 – television, satellites, microwave, computers, call centers, wireless technology – has been accompanied by determined hope that Lerner's modernization model will increase growth and productivity and produce modern cosmopolitan citizens" (Shah, 2007:24).

Ogan and her students (2009) conclude that studies have moved away from mass communication and toward information and communication technology's (ICT's) role in development, that they infrequently address development in the context of globalization and often continue to embrace a modernization paradigm despite its many criticisms:

> We believe that the more recent attention to ICTs has to do with the constant search for the magic solution to bringing information to people to transform their lives, allowing them to improve their economic condition, educate their children, increase literacy and the levels of education and spread democracy in their countries. Despite years of research that tells us that information is necessary but insufficient to bring about this change, ICTs have become the most recent iteration of the holy grail for development. And even if communication scholars know better because critical scholarship written over the last 30 years has told them so, newcomers to this field from other information-based disciplines may not have such close acquaintance with that literature. Furthermore, because of the appeal of the modernization paradigm, there is a tendency to forget that it cannot work.
>
> (Ogan et al, 2009: 667–8)

The findings by Jo Ellen Fair, Hemant Shah and Christine Ogan present us with a clear, but at the same time complex picture of the CSSC field. The implicit assumptions on which the so-called dominant modernization paradigm are built still linger on and continue to influence the policy making and planning discourse of major actors in the field of CSSC&D, both at the theoretical and applied levels.

Still to come

The uniqueness of this book lays in three original contributions. First, it puts forth a vision of communication, sustainability and development and social change that synthesizes current perspectives in both research and practice in the field. This conceptual vision is rooted in recent theoretical developments and applied programs that invite us to rethink conventional models and the boundaries of conceptual debates about development, moving the discussion into areas such as social change and social justice.

Second, it brings together both academic and programmatic perspectives to discuss a common set of questions focused on the role of communication for social change, particularly on global sustainability-related issues. The book aims to foster a conversation between academics and practitioners around questions of common interest that can inform further theoretical refining and programmatic experiences.

Third, the book covers a wide range of issues in communication (i.e., community mobilization, information technologies, gender and social norms, mass media, interpersonal communication, integrated communication approaches) and sustainability from a development and social change perspective, drawing from a variety of communication theories as well as experiences of various projects from around Asia and Africa.

I believe that presenting both theoretical and programmatic diversity is necessary to draw solid and nuanced conclusions and meaningful recommendations.

The second chapter, by Kosta Kostadinov and Jagadish Thaker, establishes the obvious links between SD and climate change. It summarizes the recommendations of the 1992 Rio Summit, and evaluates the results in view of the most recent Rio+20 Summit in June 2012. The authors argue that the mainstreaming of climate change into SD has had a beneficial effect on climate change talks. For that reason, the Rio+20 Summit was considered an important juncture to centralize the issue of climate change within the broader context of SD.

Chin Saik Yoon observes that past successes in communication for change were achieved because development initiatives offered tangible returns to communities who participated – these included higher income, improved health and enhanced quality of life. However, the solutions to climate change require us to consume less and adopt ways of living that may seem austere compared to what we have enjoyed in the past. He is therefore more cautious and warns that the long time span required to make solutions work – some of it projected to take a couple of generations – needs to be considered carefully. Communities need to adopt altruistic values and actions, which communicators are ill prepared to promote, facilitate and participate in, to address these emerging climate change problems.

Boonlert Supadhiloke synthesizes Bhutan's Gross National Happiness and Thailand's Sufficiency Economy philosophies as an alternative paradigm for SD. He explains that both development philosophies are based on Buddhist doctrine and are geared toward the goal of human happiness, rather than economic growth. He introduces the communication component known as "right speech" as an integral part of the development paradigm. A "right speech" guarantees an individual's rights to freedom of expression and information, which are basic human rights and reflect such ethical values as truthfulness and morality.

Kiran Prasad highlights the role women have played in environmental conservation in the developing world. In India, the struggle for natural resources and sustainable agriculture, and the campaign for the right to information, have been led by women. As women are often excluded from the mainstream media, they need to make use of alternative media and other forms of social mobilization. Prasad argues that these alternative forms of green communication have resulted in a paradigm shift in SD.

The central argument in Kamga and Cishahayo's chapter is that the debate regarding ICT for development should now be engaged within the framework of proportional technologies. By proportional technologies, they mean technological practices born in a given context and consistent with other established practices of the said context. Such a framework rests on the assumption that it is through practices that social changes occur. This involves, they contend, appreciating local needs, cultural values and technical competences, as well as the evolving character of the African society.

Royal Colle and Tran Van Dien start from the observation that two important movements have substantially influenced social change initiatives in the developing nations over the past couple of decades.

The first is the growing importance of information and communication technologies (ICT), fueled especially by the emergence of more affordable computers, mobile phones and wireless Internet. Their potential lies in the recognition by development agencies, including both government and non-government organizations, that ICTs can play a major role in reaching targets such as the Millennium Development Goals by 2015. A challenge is to bring these assets into the lives of those who can benefit from them.

A second movement is the attempt to reinvent higher education institutions to make them more relevant to the world around them, and to include them among the tools for SD. This chapter brings these movements together to explore the role that universities can play in nurturing the establishment of public access centers (or telecenters) – places where disadvantaged people can join the information society and improve their well-being.

The Indian state of Orissa has seen huge changes in its climate and environment with recurring floods and droughts becoming commonplace. In recent years the state has looked to industrialize quickly by opening up its rich natural resources to national and international industrial houses. Most mineral resources are located in interior areas of Orissa, which are ecologically pristine and inhabited by several tribes, for whom lands are sacred. Opening up the state's natural resources not only has environmental repercussions, but has led to the physical and spiritual dislocation of tribal communities. Maitreyee Mishra draws on three case studies that look at people's movements against displacement and environmental change in Orissa, while also examining the role of mediated and unmediated platforms in communicating about the environment. She poses some important questions on the ecological cost of development.

Equally important questions of a more socio-cultural nature are raised by Patchanee Malikhao in her assessment of the sustainability of sexual behavior among Thai students in Bangkok. The reported changes in sexual contacts among Thai youth over the past decade have been explained as consequences of "globalization" and the Western cultural invasion of Thai cultural values. Malikhao adopts a social constructionist approach to explain how the infusion of Thai and Western culture, via the use of the mass media and new media, has affected the lives and worldviews of young Bangkokians. The question is raised as to whether the resulting hybridity is sustainable in the long run.

Henri-Count Evens presents a rather bleak picture of the role of the media in reporting climate change in Africa and especially Zimbabwe. He argues that the majority of people in Africa are increasingly ignorant citizens who are not familiar with the climate change story, because the mainstream newsmedia are structurally, professionally and culturally not interested nor capable of reporting environmental stories properly. In short, eco-phobia is still rampant in most newsrooms.

The concluding chapter reiterates the question of sustainability in communication for development and social change. It argues once again that sustainable social change can only be achieved in combination with and incorporating aspects of the wider environment that influences (and constrains) structural and sustainable change.

As a consequence of globalization, media liberalization, rapid economic and social change, and the emergence of new ICTs, communication for SD is facing old and new challenges, such as poverty reduction, the promotion of food security, rural development and sustainable livelihood, the empowerment of women, girls and senior citizens, a good and affordable health care system for all.

The rise of the South

Coincidentally, while we were preparing this book for final submission, the freshly published 2013 Human Development Report (Malik, 2013) further substantiated the gist of the arguments developed in this book with additional empirical evidence. The 2013 Human Development Report identifies four specific areas of focus for sustaining development momentum: *enhancing equity*, including on the gender dimension; enabling greater voice and *participation of citizens*, including youth; *confronting environmental pressures*; and *managing demographic change*.

For the first time in 150 years, the combined output of the developing world's three leading economies – Brazil, China and India – is about equal to the combined gross domestic products (GDPs) of the longstanding industrial powers of the North: Canada, France, Germany, Italy, the United Kingdom and the United States. This represents a dramatic rebalancing of global economic power.

The middle class in the South is growing rapidly in size, income and expectations. The South is now emerging alongside the North as a breeding ground for technical innovation and creative entrepreneurship. It is not only the larger countries – notably, Brazil, China, India, Indonesia, Mexico, South Africa, Thailand and Turkey – that have made rapid advances, substantial progress has also been made in smaller economies, such as Bangladesh, Chile, Ghana, Mauritius, Rwanda and Tunisia.

However, "unless people can participate meaningfully in the events and processes that shape their lives, national human development paths will be neither desirable nor sustainable", the report claims (Malik, 2013: 18). In that regard, educating women through adulthood is the closest thing to a "silver bullet" formula for accelerating human development. This argument is made in several chapters of this book, especially in the chapter by Kiran Prasad.

While we welcome the "rise of the South", we should ask ourselves whether the glass is half empty or half full. In any case, this book proves that there still remains some way to go.

Acknowledgements

Most of the chapters in this volume were first presented at two International Conferences which form part of Orbicom, the United Nations Educational, Scientific and Cultural Organization (UNESCO) chairs in communication network's, global project *Future Imperatives of Development Communication and Information*. The aim of this project is to focus the attention of the most experienced and innovative information and communication scholars, practitioners and policy makers on the new challenges in world development and sustainability. The first conference took place in Brussels (25 May 2010) at the headquarters of Signis, the World Catholic Association for Communication. The second conference was held at the United Nations Conference Center in Bangkok, Thailand, from 20 to 22 December 2010. This conference was jointly coordinated by the Faculty of Journalism and Mass Communication at Thammasat University; the Department of Communication at the University of Nagasaki; and the Communication for Sustainable Social Change center (CSSC) at UMass Amherst.

We wish to acknowledge the assistance and support provided by the above-mentioned organizations who contributed to the success of these conferences and indirectly also made this book possible. We also wish to thank Felicity Plester, Senior Commissioning Editor for Film, Culture and Media Studies, Chris Penfold, Editorial Assistant, at Palgrave Macmillan, and Sunita Jayachandran and her team at Integra, India, for their editorial support and assistance.

Notes

1. As this book addresses the issue of SD and green communication from a variety of disciplinary perspectives, thematic explorations and multi-dimensional levels, we have attempted to gather a selective sample of relevant sources in the basic bibliography for further reading and more details.
2. For more details, see reports in *The Economist*, April to June 2004, or visit *http://www.copenhagenconsensus.com*

Basic bibliography

Alfaro Moreno, Rosa María (2006). *Otra brújula: Innovaciones en comunicación y desarrollo*. Lima, Peru: Asociación de Comunicadores Sociales Calandria.

Alvarez Nobell, Alejandro (2011). *Measurement and evaluation in communication* (in Spanish). Malaga: Institute for Research in Public Relations (IIRP).

Amin, Samir (2006). The millennium development goals: A critique from the south. *Monthly Review* 57(10). http://www.monthlyreview.org/0306amin.htm.

Anderson, B. (1983). *Imagined communities: Reflections on the origin and spread of nationalism*. London: Verso.

Anyaegbunam, C., Mefalopulos, P. & Moetsabi, T. (2004). *Participatory rural communication appraisal: Starting with the people*. Rome: Food and Agriculture Organization of the United Nations (FAO). http://www.fao.org/sd/dim_kn1/docs/y5793e00.pdf

Bamberger, M. (2009). *Institutionalizing impact evaluation within the framework of a monitoring and evaluation system*. Washington, DC: World Bank.

Bamberger, M., Rao, V. & Woolcock, M. (2010). *Using mixed methods in monitoring and evaluation: Experiences from international development*. Manchester: The World Bank Development Research Group.

Barbero, J.-M. (1993). *Communication, culture and hegemony, from the media to mediations*. London: Sage.

Bauman, Z. (1998). *Globalization: The human consequences*. New York: Columbia University Press.

Baumann, G. (1999). *The multicultural riddle. Rethinking national, ethnic, and religious identities*. London: Routledge.

Bayart, J.-F. (2004). *Le gouvernement du monde*. Paris: Fayard.

Becker, D. R. (2002). *A participatory approach to social impact assessment: The interactive community forum*. PhD College of Graduate Studies. Idaho: University of Idaho.

Beltran, L. R. (1976). TV Etchings in the minds of Latin Americans: Conservatism, materialism and conformism, Paper presented to the IAMCR Conference, Leicester, 30 August–3 September.

Bennett, F. & Roche, C. (2000). Developing indicators: The scope for participatory approaches, *New Economy*, 4, 24–28.

Berger, P. & Huntington, S. (eds.) (2002). *Many globalizations: Cultural diversity in the contemporary world*. New York: Oxford University Press.

Besette, G. 2004. *Involving the community: A guide to participatory Development communication*. Penang: Southbound/IDRC.

Bhambra, G.K. 2007. *Rethinking modernity. Postcolonialism and the sociological imagination*. London: Palgrave Macmillan.

Blattman, Chris. (2008). *MDGS: Planning for failure*. http://chrisblattman.com/2008/02/11/mdgs-planning-for-failure/.

Blewitt, J. (ed.) (2008). *Community, empowerment and sustainable development*. Totnes UK: Green Books.

Booth, D. & Lucas, H. (2002). *Good practice in the development of PRSP indicators and monitoring systems*. London: Overseas Development Institute.

Bordenave, J. D. (1977). *Communication and rural development*. Paris: UNESCO.

Boyd-Barrett, O. (1977). Media imperialism: Towards an international framework for the analysis of media systems. In J. Curran, M. Gurevitch & J. Woollacott (eds.), *Mass communication and society*. London: Arnold.

Braman, S. & Sreberny-Mohammadi, A. (eds). (1996). *Globalization, communication and transnational civil society*. Cresskill, NJ: Hampton Press.

Burgess, J. 2010. *Evaluating the evaluators: Media freedom indexes and what they measure*. Washington, DC: National Endowment for Democracy.

Byrne, A., Gray-Felder, D., Hunt, J. & Parks, W. (2005). *Measuring change: A guide to participatory monitoring and evaluation of communication for social change*. Orange, NJ: Communication for Social change Consortium. http://www.communicationforsocialchange.org/publications-resources.php?id=282

Byrne, Ailish. (2008). *Working toward evidence-based process: Evaluation that matters*. www.communicationforsocialchange.org/mazi-article, accessed 11 December 2010.

C-Change (2010). *A learning package for Social and Behavior Change Communication (SBCC)*. www.c-changeprogram.org

Canclini, N. G. (1993). *Culturas hibridas. Estrategias para entrar y salir de la modernidad*. Grijalbo.

Cardoso, F. H. & Falletto, E. (1969). *Dependencia y desarrollo en América Latina*. Mexico D.F.: Siglo XXI.

Castaneda, M. & Alfaro, R. M. (eds). (2003). *Relaciones entre Estado y Sociedad Civil. Concertación o vigilancia?*, Lima: Calandria.

Catley, A., Burns, J., Abebe, D. & Suji, O. (2007). *Participatory impact assessment: A guide for practitioners*. Medford: Feinstein International Centre, Tufts University.

Chambers, R. (2005). *Ideas for development*. London: Earthscan.

Chambers, R. (2008). *Revolutions in development inquiry*. London: Earthscan.

Chen, Guo-jie. (2001). Sustainable development facing challenges in 21st century. *Chinese Georgraphical Science*, 11(4), 289–293.

Chew, S. C. & Denemark, R. A. (eds.) (1996). *The underdevelopment of development: Essays in honor of Andre Gunder Frank*. Newbury Park, CA: Sage.

Childers, E. & Vajrathon, M. (1968). *Development support communication for project support*. Bangkok: UNDP.

Cimadevilla, G. & Carniglia, E. (eds.) (2004). *Comunicacion, Ruralidad y Desarrollo. Mitos, paradigmas y dispositivos del cambio*. Buenos Aires, Argentina: Instituto Nacional de Tecnologia Agropecuaria.

Confino, J. (2011). Is economic growth incompatible with sustainable development?, *The Guardian*, 7 September 2011. www.guardian.co.uk/sustainable-business/ (accessed on 8 September 2011).

Danida. (2005). *Monitoring and indicators for communication for development*. In T. A. Service (ed.). Copenhagen: Royal Danish Ministry of Foreign Affairs. http://webzone.k3.mah.se/projects/comdev/_comdev_PDF_doc/Danida_ComDevt.pdf

De Cuéllar, J. P. (ed.) (1995). *Our creative diversity. Report of the World Commission on culture and development*. Paris: UNESCO.

De La Court, T. (1990). *Beyond Brundtland: Green development in the 1990s*. London: Zed Books.

de Sousa Santos, B. (ed.) (2007). *Democratizing democracy: Beyond the liberal democratic canon*. London: Verso.

Delcourt, L. (2009). Retour de l'Etat. Pour quelles politiques sociales?, *Alternatives Sud* XVI(2), Louvain-la-Neuve, Belgium: Centre Tricontinental.

Deruyttere, A. (ed.) (2006). *Operational policy on indigenous peoples and strategy for indigenous development.* Washington, DC: Inter-American Development Bank.

DFID (2001). *Sustainable livelihoods guidance sheets, section 2: Framework.* London: Department of International Development, 26pp. www.livelihoods.org/onfo/info_guidanceSheets.html

Dowmunt, T. (ed.). (1993). *Channels of resistance. Global television and local empowerment.* London: BFI.

Downing, J. (2001). *Radical media: Rebellious communication and social movements.* Thousand Oaks, CA: Sage.

Eide, E. & Kunelius, R. (eds.) (2012). *Media meets climate. The global challenge for journalism.* Goteborg: Nordicom.

Ekins, P. (1993). "Limits to growth" and "sustainable development": grappling with ecological realities, *Ecological Economics*, 8, 269–288.

Elliott, A. & Lemert, C. (2006). *The new individualism: The emotional costs of globalization.* London: Routlege.

Elliott, J. (1994). *An introduction to sustainable development.* London: Routledge.

Enghel, F. & Wilkins, K. (eds.) (2012). *Communication, media and development.* Goteborg: Nordicom.

Escobar, A. (2008). *Territories of difference. Place, movements, life, redes.* Durham: Duke University Press.

Esteva, G. & Prakah, M. S. (1998). *Grassroots post-modernism: Remaking the soil of cultures.* London: Zed Books.

Estrella, M. (ed.) (2000). *Learning from change: Issues and experiences in participatory monitoring and evaluation.* London: Intermediate Technology Publications. http://www.idrc.ca/openebooks/895-3/#page_201

Fair, J. E. (1988). *A meta-research of mass media effects on audiences in developing countries from 1958 through 1986.* Unpublished doctoral dissertation. Bloomington, IN: Indiana University.

Fair, J. E. (1989). 29 years of theory and research on media and development: The dominant paradigm impact. *Gazette,* 44, 129–150.

Fair, J. E. & Shah, H. (1997). Continuities and discontinuities in communication and development research since 1958. *Journal of International Communication* 4(2), 3–23.

Fals Borda. (1991). *Knowledge and social movements.* Santa Cruz, CA: Merrill Publications.

Featherstone, M., ed. (1990). *Global culture: Nationalism, globalization and modernity.* London: Sage.

Figueroa, Maria Elena, Kincaid, D. Lawrence, Rani, Manju & Lewis, Gary. (2002). *Communication for social change: An integrated model for measuring the process and its outcomes,* Communication for Social Change Working Paper Series, No.1. New York NY: The Rockefeller Foundation.

Flint, W. (2007). *Sustainability Manifesto.* Exploring sustainability: Getting inside the concept. http://www.eeeee.net/sd_manifesto.htm, Accessed on 17 December 2011.

Foucault, M. (1982). The subject and power. *Critical Inquiry,* 8(4), 777–795.

Fowler, A. (2003). *International development frameworks, policies, priorities and implications. A basic guide for NGOs.* Oxfam Canada/Oxfam Quebec: DLR International.
Fraser, C. & Restrepo-Estrada, S. (1992). *Putting politicians under pressure: A case study of advocacy and social mobilization for children linked to decentralization and elections in Colombia.* Rome: Food and Agriculture Organization.
Fraser, C. & Restrepo-Estrada, S. (1998). *Communicating for development: Human change for survival.* London: I. B. Tauris Publishers.
Freire, P. (1983). *Pedagogy of the oppressed.* New York: Continuum.
Friedmann, J. (1992). *Empowerment: The politics of alternative development.* Cambridge, MA: Blackwell.
Fuglesang, Andreas (1982). *About understanding: Ideas and observations on cross-cultural communication.* Uppsala, Sweden: Dag Hammarskjöld Foundation.
Gawor, Leszek (2008). Globalization and its alternatives: Antiglobalism, alterglobalism and the idea of sustainable development. *Sustainable Development*, 16, 126–134.
Geertz, C. (1973). *The interpretation of cultures.* New York: Basic Books.
Geertz, C. (1983). *Local knowledge: Further essays in interpretative anthropology.* New York: Basic Books.
Giddens, A. (1995). *Modernity and self-identity: Self and society in the late modern age.* Cambridge: Polity.
Gosling, L. & M. Edwards (2003). Impact assessment. In L. Gosling & M. Edwards (eds.), *Toolkits: A practical guide to monitoring, evaluation, impact assessment* (pp. 125–142). London: Save the Children Fund.
Grillo, M., Berti, S. & Rizzo, A. (1998). *Discursos locales.* Rio Cuarto: Universidad Nacional de Rio Cuarto.
Guijt, I. and P. Sidersky (1996). Agreeing on Indicators. *ILEIA Newsletter.* 12, 9–11.
Hafez, K. (2007). *The myth of media globalization.* Cambridge, UK: Polity Press.
Hall, S. (ed.) (1997). *Culture, media and identities.* London: Sage.
Hall, Stuart & du Gay, Paul (eds.) (1996). *Questions of cultural identity.* London: SAGE.
Hannerz, U. (1996). *Transnational connections: Culture, people, places.* London: Routledge.
Harindranath, R. (2006). *Perspectives on global cultures.* Maidenhead: Open University Press.
Harrison, L. (1985). *Underdevelopment is a state of mind. The Latin American case* (2nd ed.). Lanham, MD: Madison Books.
Hass, Larry, Leonardo, Mazzei & O'Leary, Donal (2007). *Setting standards for communication and governance. The example of infrastructure projects.* World Bank Working Paper No 121. Washington, DC: World Bank.
Held, D. (ed.) (2000). *A globalizing world? Culture, economic, politics.* London: Routledge.
Held, David & McGrew, Anthony (2007). *Globalization/anti-globalization: Beyond the great divide.* 2d ed. Cambridge, UK: Polity.
Hemer, O. & Tufte, T. (Eds). (2006). *Media and glocal change. Rethinking communication for development*: Gotebeorg: NORDICOM, Clacso Books.
Hirst, P. & Thompson, G. (1996). *Globalization in question: The international economy and the possibilities of governance.* London: Polity.

Hoffmann, V. (2007). Five editions (1962–2003) of Everett Rogers's diffusion of innovations. *Journal of Agricultural Education and Extension*, 13(2), 147–158.

Holte-McKenzie, M., Forde, S. et al. (2006). Development of a participatory monitoring and evaluation strategy. *Evaluation and Program Planning*, 29, 365–376.

Hopper, P. (2007). *Understanding cultural globalization*. Cambridge: Polity.

Horbach, Jens. (2005). *Methodological Aspects of an Indicator System for Sustainable Innovation*. Sustainability and Innovation book series, Physica-Verlag HD.

Hospes, Otto. (2008). Evaluation evolution?, *The Broker* 8 (June), 24–26. www.thebrokeronline.eu

Hull, Zbigniew. (2008). Sustainable development: Premises, understanding and prospects. *Sustainable Development*, 16, 3–80.

Inagaki, N. (2007). *Communicating the impact of communication for development. Recent trends in empirical research*. Washington, DC: The World Bank, Working Paper Series, Development Communication Division.

Jacobson, T. & Servaes, J. (Eds.) (1999). *Theoretical approaches to participatory communication*. Cresskill, NJ: Hampton.

James, J. (2013). *Digital interactions in developing countries. An economic perspective*. New York: Routledge.

Jameson, F. & Miyoshi, M. (eds.) (1998). *The cultures of globalization*. Durham: Duke University Press.

Janssens, M., Bechtoldt, M., de Ruijter, A., Pinelli, D., Prarolo, G. & Stenius, V. (eds.) (2010). *The sustainability of cultural diversity. Nations, cities and organizations*. Cheltenham: Edward Elgar.

Johannesburg Summit 2002, 26 August–4 September, 2002, http://www.un.org/jsummit/html/basic_info/basicinfo.html, accessed 1 October, 2009.

Johnston, H. & Klandermans, B. (eds.) (1995). *Social movements and culture*. London: University College Press.

Kennedy, T. (2008). *Where the rivers meet the sky. A collaborative approach to participatory development*. Penang: Southbound.

Khampa (2009). *Participatory development communication and people's happiness in a Bhutanese community*, M.A. Thesis at the Graduate School of Bangkok University, Bangkok, Thailand.

Knapp, M. & Daly, J. (eds) (2002). *Handbook of interpersonal communication*. London: Sage.

Kronenburg, J. (1986). *Empowerment of the poor: A comparative analysis of two development endeavors in Kenya*. Nijmegen: Third World Center.

Laswell, H. (1948). The structure and function of communication in society. In L. Bryson (ed.). *The communication of ideas*. New York: Harper & Brothers.

Lele, S. (1991). Sustainabile development: A critical review. *World Development* 19(6), 607–621.

Lennie, June & Tacchi, Jo (2010). *Evaluating communication for development: Trends, challenges and approaches. Draft report on a literature review and consultations conducted for the project: UN Inter-agency research, monitoring and evaluation resource pack for communication for development programs*. New York: UNICEF, 3 December 2010.

Lennie, June & Tacchi, Jo (2013). *Evaluating communication for development: A framework for social change*. London: Earthscan/Routledge.

Lerner, D. (1958). *The passing of traditional society: Modernizing the Middle East.* New York: Free Press.

Lerner, D. (1977). Communication and development. In D. Lerner & L. Nelson (eds.), *Communication research: A half-century appraisal.* Honolulu: East-West Center.

Lerner, Daniel & Schramm, Wilbur (Eds.) (1967). *Communication and change in the developing countries.* Honolulu: East-West Center.

Lie, R. (1998). What's new about cultural globalization? Linking the global from within the local. In J. Servaes & R. Lie (eds.) *Media and politics in transition: Cultural identity in the age of globalization* (pp. 141–155). Leuven: ACCO.

Lie, R. (2003). *Spaces of intercultural communication: An interdisciplinary introduction to communication, culture, and globalizing/localizing identities.* Cresskill, NJ: Hampton Press.

Lie, R. & Servaes, J. (2000). Globalization: consumption and identity. In G. Wang, J. Servaes & A. Goonasekera (eds), *The new communications landscape: Demystifying media globalization.* London: Routledge.

Loomis, T. (2000). Indigenous populations and sustainable development: Building on indigenous approaches to holistic, self-determined development. *World Development,* 28(5), 893–910.

MacBride, S., ed. (1980). *Many voices, one world: Communication and society, today and tomorrow.* Paris: UNESCO.

Malik, K. (ed.) (2013). *The rise of the south: Human progress in a diverse world. Human development report.* New York: UNDP.

Mannberg, Mariann & Wihlborg, E. (2008). Communicative planning- Friend or Foe? Obstacles and opportunities for implementing sustainable development locally, *Sustainable Development,* 16: 35–43.

Mansell, R. & When, U. (eds). (1998). *Knowledge societies: Information technology for sustainable development.* Oxford: Oxford University Press.

Marling, W. (2006). *How "American" is globalization?* Baltimore: John Hopkins University Press.

Martinez-Gomez, R. & Lubethin, M. (eds). (2008). *Comunicacion y desarrollo: en busca de la coherencia.* Madrid: Fundacion Carolina y Siglo XXI.

Mato, D. (2001). *Globalización, cultura y transformaciones sociales.* Buenos Aires: Consejo Latinoamericano de Ciencias Sociales.

Mato, D. (2003). *Políticas y Identidades y Diferencias Sociales.* Caracas: FACES-UVC.

Mattelart, Armand (2007). *Diversite culturelle et mondialisation.* Paris: La Decouverte.

Max-Neef, Manfred (1991). *Human scale development. Conception, application and further reflections.* New York: The Apex Press.

McKee, N., Manoncourt, E., Saik Yoon, C. & Carnegie, R. (2000). *Involving people, evolving behaviour.* Penang: Southbound and UNICEF.

McKee, N., Manoncourt, E., Saik Yoon, C. & Carnegie, R. (2003). Involving people, evolving behaviour. In J. Servaes (ed.), *Approaches to development: Studies on communication for development.* Paris: UNESCO.

McMichael, P. (2008). *Development and social change: A global perspective.* Los Angeles: Pine Forge Press.

McQuail, D. (1983). *Mass communication theory.* London: Sage.

McQuail, D. (2005). *Mass communication theory,* 5th ed. London: Sage.
Melkote, S.R. & Steeves, H.L. 2001. *Communication for development in the third world: Theory and practice for empowerment.* London: Sage Publications.
Millenium Development Goals Report. (2009). New York http://www.un.org/millenniumgoals/pdf/MDG_Report_2009_ENG.pdf, accessed 1 October, 2009.
Mitchell, Paul & Chaman-Ruiz, Karla (2007). *Communication-based assessment for bank operations,* World Bank Working Paper No 119, Washington DC: World Bank.
Mody, B. (ed.) (1997). Communication and development: Beyond panaceas. *The Journal of International Communication,* 4(2), 1–138.
Mody, Bella (ed.) (2003). *International and development communication: A 21st-century perspective.* Thousand Oaks, CA: Sage.
Mozammel, M. & Schechter, G. (2005). *Strategic communication for community-driven development.* Washington DC: Worldbank.
NASA (2012). NASA Finds 2011 Ninth-Warmest Year on Record, http://www.nasa.gov/topics/earth/features/2011-temps.html, 01.19.12 (accessed on 25 January 2012).
Nash, J. (ed.) (2005). *Social movements: An anthropological reader.* Masden, MA: Blackwell.
Nederveen Pieterse, J. (2010). *Development theory (2nd edition).* Los Angeles: Sage.
Nerfin, Marc (ed.) (1977). *Another development. Approaches and strategies,* Uppsala: Dag Hammerskjold Foundation.
Nyamu-Musembi, C. & Cornwall, A. (Nov 2004), *What is the "rights-based approach" all about? Perspectives from international development agencies.* IDS Working Paper 234. Brighton: Institute of Development Studies.
O'Sullivan, G.A., Yonkler, J.A., Morgan, W. & Merritt, A.P. (2003). *A Field guide to designing a health communication strategy.* Baltimore, MD: Johns Hopkins Bloomberg School of Public Health/Center for Communication Programs. http://www.jhuccp.org/node/1033om
Offenheiser, R.C. & Holcombe, S. (2001). *Challenges and opportunities of implementing a rights-based approach to development: an Oxfam America perspective.* Paper presented at the Conference on Northern Relief and Development NGOs: New Directions in Poverty Alleviation and Global Leadership Transitions (2–4 July). Oxford: Balliol College.
Ogan, C. L., Bashir, M., Camaj, L., Luo, Y., Gaddie, B., Pennington, R., Rana, S. & Salih, M. (2009). Development communication: The state of research in an era of ICTs and globalization. *The International Communication Gazette,* 71(8), 655–670.
Omoto, A. (2005). *Processes of community change and social action.* Mahwah, NJ: Lawrence Erlbaum.
Papa, M.J., Singhal, A. & Papa, W. 2006. *Organizing for social change. A dialectic journey of theory and praxis.* New Delhi: Sage Publications.
Parks, Will, Gray-Felder Denise, Hunt, Jim & Byrne Ailish. (2005). *Who measures change? An introduction to participatory monitoring and evaluation for communication for social change.* Orange, NJ: CFSC Consortium.
Payutto, P. (1998). *Sustainable development.* Bangkok: Buddhadham Foundation.
Puddephatt, A. (2007). *Defining indicators of media development. Background paper.* Paris: IPDC and UNESCO.

Puddephatt, A., Horsewell, R. & Menheneott, G. (2009). *Discussion paper on the monitoring and evaluation of UN-assisted communication for development programmes. Recommendations for best practice methodologies and indicators.* Paper presented at the 11th UN inter-agency round table on communication for development. 11th–13th March. Washington DC: UNDP–World Bank.

Rangel, C. (1977). *The Latin Americans: Their love-hate relationship with the United States.* New York: Harcourt Brace Jovanovich.

Reardon, C. (2003). *Talking cure: A case study in communication for social change.* Working Paper Series. Orange, NJ: Communication for Social Change.

Regeer, B., Hoes, A., Sanne, M., Caron-Flinterman, F. & Bunders, J. (2009). Six guiding principles for evaluating mode-2 strategies for sustainable development. *American Journal of Evaluation,* 30(4), 515–537.

Reynolds, Godfrey Johnson & Tommy, John (2002). *Environmental Assessment Final Report of the Health Sector Reconstruction and Development Project for Republic of Sierra Leone,* Report No: E1630, World Bank Reports and Documents, 29 October 2002. Washington DC: World Bank.

Robertson, R. (1992). *Globalization: Social theory and global culture.* London: Sage.

Rockefeller Foundation. (1999). *Communication for social change: A position paper and conference report.* New York NY: Rockerfeller Foundation.

Rockefeller Foundation. (2002). *Pocantico statement on the need for national indicators of sustainability for the United States of America* (pp. 1–6). New York, NY: Rockerfeller Foundation.

Rogers, E. M. (1962). *The diffusion of innovations.* New York: The Free Press.

Rogers, E. M. (1983). *Diffusion of innovations.* 3rd ed. New York: Free Press.

Rogers, E. M. (1986). *Communication technology: The new media in society.* New York: The Free Press.

Rogers, E.M. (2003). *Diffusion of innovations.* New York: Free Press.

Rogers, Everett, M. (1976). Communication and development: The passing of the dominant paradigm. *Communication Research,* 3(2), 213–240.

Salas, Maria A. and Tillmann, Hermann (2010). *Participatory action research: Embracing the knowledge perspective within field research.* Chiang Mai: Regional Center for Social Science and Sustainable Development.

Sanyawiwat, S. (2003). *Theories and social development strategies.* Bangkok: Chualalongkorn Publishing House.

Schiller, H. I. (1976). *Communication and cultural domination.* New York: International Arts and Sciences Press.

Scholte, J. (2005). *Globalization: A critical introduction.* New York: Palgrave.

Schramm, W. (1964). *Mass media and national development: The role of information in the developing countries.* Stanford, CA: Stanford University Press.

Schramm, Wilbur & Daniel Lerner (eds.) (1976). *Communication and change: The last ten years – and the next.* Honolulu: University Press of Hawaii.

Sen, A. (2004). Cultural liberty and human development. In S. Fukuda-Parr (ed.), *Human development report: Cultural liberty in today's diverse world.* New York: United Nations Development Programme.

Servaes, J., Jacobson, T. & White, S. (eds.) (1996). *Participatory communication for social change.* New Delhi: Sage, 287 pp.

Servaes, J. & Lie, R. (eds.) (1998). *Media and politics in transition: Cultural identity in the age of globalization.* Louvain-la-neuve: Acco.

Servaes, J. (1999). *Communication for development: One world, multiple cultures.* Cresskill, NJ: Hampton Press.
Servaes, J. (2000). Advocacy strategies for development communication. In J. Servaes (ed.). *Walking on the other side of the information highway: Communication, culture and development in the 21st century* (pp. 103–118). Penang: Southbound.
Servaes, J. (ed.) (2003). *Approaches to development: Studies on communication for development.* Paris: UNESCO.
Servaes, J. & Liu, S. (eds.) (2007). *Moving targets: Mapping the paths between communication, technology and social change in communities.* Penang: Southbound.
Servaes, Jan (2007). Harnessing the UN system into a common approach on communication for development. *The International Communication Gazette*, 69(6) 483–507.
Servaes, Jan (ed.) (2007). Communication for development. Making a difference – A WCCD background study. In *World Congress on Communication for Development: Lessons, Challenges and the Way Forward*, World Bank, Washington, DC, 209–292.
Servaes, Jan & Malikhao, Patchanee (2007a). *Communication and sustainable development*, FAO, Rome, pp. 1–38.
Servaes, Jan & Malikhao, Patchanee (2007b) Communication and sustainable development. In J. Servaes & S. Liu, (eds.), *Moving targets. Mapping the paths between communication, technology and social change in communities* (pp. 11–42). Southbound, Penang.
Servaes, Jan (ed.) (2008). *Communication for development and social change.* Sage, Los Angeles-London-New Delhi-Singapore.
Servaes, Jan (2009). Communication policies, good governance and development journalism. *Communication: South African Journal for Communication Theory and Research*, 35(1), 50–80.
Servaes, Jan (2011). Social change. *Oxford Bibliographies Online* (OBO), Oxford University Press, New York, 58 pp. http://www.oxfordbibliographiesonline.com/display/id/obo-9780199756841-0063
Servaes, Jan (2012). Comparing development communication, In Frank Esser & Thomas Hanitzsch (eds.) *The handbook of comparative communication research* (pp. 64–80). New York: Routledge.
Servaes, Jan, Polk Emily, Song Shi, Danielle Reilly & Thanu Yakupitijage (2012a). Sustainability testing for development projects, *Development in Practice* 22(1), 18–30. http://www.developmentinpractice.org/
Servaes, Jan, Emily, Polk, Shi, Song, Reilly, Danielle & Yakupitijage, Thanu (2012b). Towards a framework of sustainability indicators for "communication for development and social change" Projects. *The International Communication Gazette (Sage)*, 74(2), 99–123.
Silvia, Nagy-Zekmi & Karyn, Hollis (eds.) (2012). *Global academe: Engaging intellectual discourse*. New York: Palgrave-MacMillan.
Shah, Anup. (2005). *Sustainable development introduction.* www.globalissues.org, accessed 1 October 2009.
Shah, H. (2007). Meta-research of development communication studies, 1997–2005: Patterns and trends since 1958, Paper presented to ICA, San Francisco, 24–27 May.

Sivaraksa, Sulak (2010). *The wisdom of sustainability. Buddhist economics for the 21st century*. Chiang Mai: Silkworm Books.

Sklair, L. (1991). *Sociology of the global system: Social change in global perspective*. Hemel Hempstead: Harvester Wheatsheaf.

Skowrownski, Antoni. (2008). A civilization based on sustainable development: its limits and prospects. *Sustainable Development*, 16, 117–125.

Skuse, A. and Power, F. (2005). *AIDS communication*. London: Department of International Development.

Skuse, Andrew, Fildes, Joann, Tacchi, Jo, Martin, Kirsty, Baulch, Emma. (2007). *Poverty and digital inclusion: Preliminary findings of finding a voice* (pp. 1–56). Paris: UNESCO.

Solervicens, M. (Ed.) 2007. *Community radio social impact assessment. removing barriers. increasing effectiveness*. Ottawa: World Association of Community Radio Broadcasters (AMARC).

Sparks, C. (2007). *Globalization, development and the mass media*. London: Sage.

Stewart, C., Smith, C. A. & Denton, R. (2007). *Persuasion and social movements* (5th ed.). Long Grove, IL: Waveland Press.

Sunkel, O. & Fuenzalida, E. (1980). La transnacionalizacion del capitalismo y el desarrollo nacional, In Sunkel, O., Fuenzalida, E., Cardoso F. et al. (eds.), *Transnacionalizacion y dependencia*. Madrid: Cultura Hispania.

Supadhiloke, Boonlert (2010). Synergy of "Gross National Happiness" and "Sufficiency Economy" as an Imperative Paradigm for Communication and Sustainable Social Change, Paper International Conference *Future imperatives on communication and Information for development and social change*, Bangkok, 20–22 December 2010.

Tacchi, Jo, Slater, Don and Hearn, Greg. (2003). *EAR: Ethnographic action research training handbook*. Paris: UNESCO. http://ear.findingavoice.org/intro/index.html on November 29, 2009.

Taylor, Lance (1996). Sustainable development: An introduction. *World Development*, 24(2), 215–225.

Thanh, Hoang Xuan (Ed.) 2010. *Participatory poverty monitoring in rural communities in Vietnam*. London: Oxfam.

Tomlinson, J. (1999). *Globalisation and culture*. Cambridge: Polity Press.

Tran, Mark (2012). Put planet and its people at the core of sustainable development, urges report. *The Guardian*, 30 January 2012.

Tremblay, Solange (ed.) (2007). *Developpement durable et communications. Au-dela des mots, pour un veritable engagement*. Quebec: Presses de l'Universite du Quebec.

UN (2008). *Achieving sustainable development and promoting development coorperation*. New York: United Nations.

UN News Centre (2012). The High-level Panel on Global Sustainability presents its report to the Secretary-General on 30 January 2012 in Addis Ababa. http://www.un.org/gsp/report (accessed on 30 January 2012).

UNAIDS (2000). *National AIDS programmes: A guide to monitoring and evaluating*. Geneva: World Health Organisation. http://www.who.int/hiv/pub/me/pubnap/en

UNDP (2004). *Human development report 2004: Cultural liberty in today's diverse world*. New York: United Nations Development Programme and Oxford University Press.

UNDP (2010). *Human development report. The real wealth of nations: pathways to human development*. New York: UNDP.

UNDP and World Bank. (2009). *Discussion paper on the monitoring and evaluation of UN-assisted communication for development programmes*. Background Paper, 11th UN Inter-Agency Round Table on Communication for Development. Washington, DC. February 2009.

UNESCO. (2008a). *Media development indicators: A framework for assessing media development*. Paris: UNESCO. http://unesdoc.unesco.org/images/0016/001631/163102e.pdf

UNESCO. (2008b). Sustainable development: An evolving concept. *Education for sustainable development brief*. Paris: UNESCO.

UNFPA (2002). *Communication for development roundtable report of the communication for development roundtable*. Managua, Nicaragua, 26–28 November 2001. New York: United Nations Fund for Population Activities.

UNICEF (2008). *Essentials for excellence: Research, monitoring and evaluating strategic communication for behaviour and social change with special reference to the prevention and control of Avian Influenza/Pandemic Influenza*, Suva, Fiji, UNICEF Pacific Office. http://www.unicef.org/cbsc/files/Essentials_for_excellence.pdf

UNICEF. (2010). *United Nations inter-agency & experts' consultation on research, monitoring and evaluation in communication for development*, UNICEF, New York, 7–9 December 2010.

United Nations (2000). *United Nations Millennium Declaration*. http://www.ohchr.org/english/law/millennium.htlmUnited Nations Environment Programme (UNEP) (2012). *Global environmental outlook GEO5. Environment for the future we want*, NY: UNEP

United Nations Secretary-General's High-level Panel on Global Sustainability (2012). *Resilient people, resilient planet: A future worth choosing*. NY: UN.

United Nations. (1983). *Process of preparation of the environmental perspective to the Year 2000 and beyond*. General Assembly Resolution 38/161, 19 December 1983.

Van Hemelrijck, Adinda (2009). *LEAD measurement note on empowerment in rights-based programming. Implications for the work of Oxfam America*. Boston: Oxfam.

Wang, G., Servaes, J. & Goonasekera, A. (Eds.) (2000). *The new communications landscape demystifying media globalization*. London: Routledge.

Webb, D. & Elliott, L. (2002). *Learning to live: Monitoring and evaluating HIV/AIDS programmes for young people*. London: Save the Children Fund. http://www.reliefweb.int/rw/lib.nsf/db900sid/LHON-64LK6C/$file/SaveTheChildren_AIDS_ME_January_2002.pdf?openelement

Whaley, Denise, Weaver, Liz & Born, Paul (2010). *Approaches to measuring: Community Change Indicators*. Waterloo ON: Tamarack. www.tamarackcommunity.ca

WHO (2004). *National AIDS programmes: A guide to monitoring and evaluating national HIV/AIDS prevention programmes for young people*. Geneva: WHO.

Wilkins, K. G. (2000). *Redeveloping communication for social change: Theory, practice, and power*. Lanham, MD: Rowman and Littlefield Publishers.

Wilson, Mark. (2007). *At the heart of change: The role of communication in sustainable development*. London: Panos.

Wilson, R. & Dissanayake, W. (1996). *Global/local: Cultural production and the transnational imaginary*. London: Duke University Press.

Winterstein, D. (2010). *Media and power on the margins of Europe: The public negotiation of the Breton language and cultural identity*. Cresskill, NJ: Hampton Press.

Wolton, Dominique (2003). *L'Autre Mondialisation*. Paris: Flammarion.

World Commission on Environment and Development. (1987). *Our common future*. Published as Annex to General Assembly document A/42/427, Development and International Co-operation: Environment. New York: UN.

World Commission on Environment and Development. (1987). *Our common future*. Published as Annex to General Assembly document A/42/427, Development and International Co-operation: Environment. New York: UN.

World Economic Forum (WEF) (2012). *Global risks 2012* (7th edition). An Initiative of the Risk Response Network, Geneva: WEF.

Part I

Setting the Stage

1

Sustainable Development and Climate Change: Beyond the Rio + 20 Summit

Kosta Kostadinov and Jagadish Thaker[1]

The Rio + 20 Summit

The United Nations Conference on Sustainable Development (UNCSD), popularly known as Rio + 20, was held on 13–22 June 2012 in Rio de Janeiro, Brazil. It marked the 20th anniversary of the 1992 United Nations Conference on Environment and Development (UNCED), also known as the Earth Summit, and the 10th anniversary of the 2002 World Summit on Sustainable Development, held in Johannesburg, South Africa. Mandated by the United Nations (UN) General Assembly Resolution No. 64/236 (A/RES/64/236), the Rio + 20 Summit focused on two main themes:

1. a "green economy" in the context of sustainable development and poverty eradication; and
2. the institutional framework for sustainable development.

The first Rio Conference in 1992 led to several important outcomes, among them the Rio Declaration on Environment and Development. It also affirmed the concept of sustainable development as encompassing economic and social development along with environment protection to guide priorities for international negotiations. Moreover, the 1992 Rio Conference put together the Agenda 21 action plan, comprising a set of guidelines for sustainable development in the twenty-first century. The action plan brought forth the UN Framework Convention on Climate Change (UNFCCC), a treaty on climate change mitigation and adaptation, and the Convention on Biological

Diversity. With climate change issues coming to the fore of international negotiations, most popularly during the Conference of Parties (COP) 15 in Copenhagen in December 2009, which was attended by numerous heads of state, there were high expectations for the Rio + 20 Summit results.

The three objectives of the Rio + 20 Summit were to

- secure renewed political commitment for sustainable development;
- address the progress and implementation gaps of previous commitments; and
- address new and emerging challenges.

It was hailed as the largest UN-organized summit, having been attended by 130 heads of state and government, ministers and diplomats from 180 countries, and by over 50,000 attendees (Vidal, 2012).

The outcome of the Rio + 20 Summit was a document titled "The Future We Want", which renewed the political commitment to promote an economically, socially and environmentally sustainable future. Some of the important action plans in the document underlined the need to develop Sustainable Development Goals (SDGs), similar to the Millennium Development Goals set to expire in 2015, with more emphasis on protection of the environment and detailing the use of a "green economy" for sustainable development. Steps were taken toward a more comprehensive set of indicators other than the Gross Domestic Product (GDP), among others. Governments, corporations, civil society groups and others pledged more than $513 billion toward developing these initiatives for a more sustainable future (http://www.uncsd2012.org/). Nevertheless, the outcomes of the Rio + 20 Summit fell short of expectations and the negotiations ended with *no legally binding agreements*.

As with other international meetings, which see a lot of debate and finger-pointing over keywords and definitions, the Rio + 20 Summit discussions were also filled with much acrimony. The central contestation during the negotiations between developed and developing countries was the role of a "green economy" replacing the focus on sustainable development. The G77-China group, along with other developing countries, insisted that an emphasis on a "green economy" at the expense of social, economic and environmental protection would lead to the promotion of green industries while leaving behind the important aims of poverty reduction and energy access, which developing countries perceive to be among their primary concerns (Khor, 2012; Sanwal, 2011).

Similarly, they feared a "green economy" would give rise to unilateral "green" trade barriers and would become a new conditionality for international aid and finance. Many civil society organizations also raised concerns that the concept of a "green economy" would lead to the commodification of nature and the rise of "nature markets", where natural resources, such as water and forests, would become available for international trade, while local communities would be forced to withdraw their rights to such resources. Furthermore, developing countries were also concerned that a "green economy" would interfere with their sovereignty in national policy and implementation processes, a deviation from the original Rio principle of respect for individual member states' sovereignty to decide and implement their own policies commensurate with their national priorities. Finally, they also feared that the Rio + 20 Summit outcomes may ignore the core principle of differentiated responsibilities and require, for example, countries such as China and India to commit to the SDGs on a par with the developed countries (Jayaraman, Kohli & Mittal, 2012; Sanwal, 2011).

"The Future We Want" assuaged some of those fears by acknowledging the lack of progress from earlier commitments, centralizing eradication of poverty as an "indispensable requirement" for sustainable development, incorporating concerns about national sovereignty, strengthening of finance and technology transfer mechanisms, and urging countries to refrain from trade protectionism, among others. It also affirmed the principle of common but differentiated responsibility, enshrined in Principle 7 of the original Rio Declaration.

The final Rio + 20 Summit document is still lacking in several major objectives. One of the major critiques of the Summit was the *lack of a concrete, legally binding framework for a sustainable development future*; instead, deferring the process to various select committees to be negotiated later, and relying on voluntary pledges from governments and the private sector. For example, member countries agreed to set up a 30-member group to come up with SDGs to be presented at the 68th session of the UN Assembly. In particular, there were no concrete outcomes on the issues of technology and finance transfer to developing and underdeveloped countries so that they could move toward adopting SDGs without sacrificing national priorities about increased energy access and poverty reduction. In place of mandatory transfer of technology and funding, the outcome document referred to only "mutually agreed" terminology (Khor, 2012). Further, it deferred the process of the establishment of a mechanism for finance and technology transfer to 30 experts, tasked with coming up with finance proposals by 2014.

More noteworthy among the Rio + 20 Summit's achievements are:

- underscoring the importance of strengthening international environmental governance in the context of an institutional framework for sustainable development by strengthening and upgrading the role of the United Nations Environment Programme (UNEP) through increased funding; and
- establishing a high-level political forum in place of the Commission on Sustainable Development.

Although the Rio + 20 Summit failed to achieve any legally binding agreements, it survived the test of continuing the international dialogue on SDGs in times when international negotiations often seemed to end without any outcomes (Khor, 2012). The process of negotiations, however, did suggest an increasing trust deficit between the developed and the developing countries to act collectively toward a common set of goals. At the same time, the Summit also witnessed an increasing consensus between different civil society organizations, non-governmental organizations, and indigenous groups from across the world, on the need for more ambitious SDGs that emphasize environmental and social components of sustainable development, in contrast to a singular focus on "green" growth. It became increasingly clear that developing countries such as China, India, Brazil and South Africa, among others, are the main forces of economic growth in this century, consuming more resources as they transition millions of people out of poverty (Papa & Gleason, 2012). The policies such countries would adopt and the speed with which they would implement the changes not only depend on the ambitions of individual countries for "green growth", but also on other contentious issues such as technology transfer from developed to developing countries.

The discourse contests between the developed and the developing world about the role and responsibility of individual countries to contribute to common goals is not limited to sustainable development objectives alone and is also present at the annual UNFCCC Conference of Parties (COP) meetings. With climate scientists ever more certain that humankind has become the primary driver of climate change, and its impacts becoming increasingly visible, particularly in poor and developing countries, the goals of sustainable development are increasingly recognized in alignment with an effective international agreement on climate change mitigation and adaptation (Sathaye et al., 2007; Yohe et al., 2007).

Sustainable development and climate change: The case for an integrated approach

Sustainable development emerged as an important international issue in the 1980s, as international policy negotiations increasingly centered around the negative effects of existing economic development models on environmental standards by threatening to exhaust the resources of the earth (Redclift, 1987). The concept of sustainable development, often defined as "development that meets the needs of the present without compromising the ability of future generations to meet their own needs" (Brundtland Commission, 1987), seeks to align economic, social and ecological dimensions by articulating alternative development pathways. Some critics point out that the concept of sustainable development is too vague (Gibson, 1991), or fear that the stress on sustaining *development*, equated with a singular focus on economic growth, would lead to ignoring the core underlying principles of equity and social justice (e.g., Redclift, 1987; Robinson, 2004; Servaes, 2013). Others also argue that the term may be coopted by governments and businesses to make only cosmetic changes, or "greenwashing", without fundamentally shifting growth pathways (e.g., Athanasiou, 1996; Greer & Bruno, 1996; Najam, 1999; Tokar, 1997). Nevertheless, by balancing economic and environmental priorities, and the promise of building alternative development perspectives, it seems to have gained international currency among policy makers and has become central in several international negotiations (Robinson, 2004; Gawor, 2008).

According to the World Meteorological Organization, the year 2012 was one of the hottest years on record (WMO, 2012). Rising temperatures have resulted in accelerated melting of Artic sea ice to record levels, whereas many countries experienced extreme weather events such as heat waves, severe droughts and floods. Although climate scientists claim that no one extreme weather event can be directly linked to climate change, they believe that climate change has most likely increased their intensity and frequency. The effects of such disasters, however, are distributed unequally, with the poor in underdeveloped and developing countries being the most vulnerable. For example, from 1970 to 2008, over 95 per cent of deaths related to natural disasters occurred in developing countries (IPCC, 2012). Moreover, climate change is said to most likely affect yields of many crops, and increase the prevalence of infectious diseases such as malaria in tropical countries, thereby threatening the lives and livelihoods of millions of people. As a result, the effects of climate change, particularly in developing and underdeveloped countries, may not only impede their economic progress, but reverse the development trend achieved in the last decade (see UNDP 2007–2008 Report).

Climate change refers to the changes in the earth's average temperature patterns due to increased accumulation of greenhouse gases in the atmosphere, primarily as a result of human activities. According to the Intergovernmental Panel on Climate Change (IPCC) – the UN's primary scientific body that assesses climate change science and releases periodic reports – the warming of the earth's climate change is unequivocal, with "eleven of the last twelve years (1995–2006) ranked among the twelve warmest years in the instrumental record of global surface temperature (since 1850)" (IPCC, 2007). Scientists maintain that climate change will not only have short-term impacts such as increasing the intensity of natural disasters, but will also have long-term impacts on people's food, economic and livelihood security by causing shifting regional weather patterns, which will have a subsequent effect on crop yields. Further, it is expected that developing and poor countries will acutely feel the negative impacts of climate change given that large parts of their populations are dependent on climate-sensitive livelihoods, such as agriculture. According to the Stern Report (Stern, 2007), countries need to spend 1 per cent of global GDP to stop greenhouse gas emissions immediately, or risk paying between 5 per cent and 20 per cent of global GDP later for damages due to climate change.

As human activities are seen as the primary *drivers* of climate change, it seems imperative that rising climate change concerns must be addressed within the context of sustainable development efforts, which centralize the issue of human production and consumption patterns and emphasize the linkages between social, economic and environmental concerns. The increase in emissions of greenhouse gases from industries and the transportation sector, for example, has been identified as the primary cause of recent climate changes. Therefore, addressing the issue by redesigning the transportation sector to be more public-friendly, fuel-efficient and less fossil fuel dependent would require addressing the socially and culturally embedded patterns and processes of production and consumption.

It seems increasingly feasible that the current emissions of greenhouse gases will result in disastrous climatic conditions in future due to the laggard effects inherent in the Earth's climatic system. This means that the core principle of intergenerational equity in sustainable development – that the present consumption of resources does not undermine the ability of future generations to use such resources – is also directly linked to the climate change problem.

Another important attribute to sustainable development, enshrined in the UNFCCC principles, is the concept of *equity* as in equitable access and distribution of resources for the most vulnerable groups, such as the poor and rural women within and across countries (Yohe et al., 2007). Climate change negotiations, however, have often focused on equity between countries, mostly between developed and developing countries, without paying sufficient attention to equity for the poorest and most vulnerable countries, as well as equity within individual countries (Jayaraman, 2012).

Finally, it has to be pointed out that the development goals to secure sustainable economic prosperity for all are inherently tied with climate change action plans. Building environmentally friendly cities, for example, will help not only secure the SDG of increasing access to resources, but will also address climate change challenges. The most visible linkage between sustainable development and climate change is in the area of clean energy production, which is increasing access to energy for millions of people without degrading the environmental resources in the process of power generation.

The concept of sustainable development as comprising economic, social and environmental issues was affirmed at the 1992 Rio Summit, when climate change was still a nascent issue. Yet the 1992 Rio Summit led to the establishment of the UNFCCC, the UN's primary intergovernmental body that convenes annual meetings between 193 countries to seek policy measures for climate change mitigation and adaption. In the 20 years since then, climate change has turned into one of the biggest challenges humanity faces in the twenty-first century. For that reason, the Rio + 20 Summit was considered an important juncture to centralize the issue of climate change within the broader context of sustainable development.

The Rio + 20 Summit's report, "The Future We Want," acknowledges that the negative impacts of climate change will undermine efforts for sustainable development, particularly for developing and poor countries. Further, it recognizes that policies such as improving energy efficiency and increasing the use of renewable energy are important for sustainable development as well as for mitigating the effects of climate change (see Sathaye et al., 2007). Climate change and sustainable development are interlinked within the context of sustainable energy, where, on the one hand, policies are adopted to reduce the emissions of greenhouse gases from energy production and consumption and, on the other, access to clean energy is provided to millions who still lack access to any form of modern energy sources. Similarly, the document calls for increased measures to address biodiversity conservation, ocean acidification, desertification, deforestation, among others – issues that are closely tied with climate change.

Just as energy-efficient and renewable energy sources are important to mitigate the effects of climate change and promote sustainable development, so is the issue of *adaptation*. Adaptation to climate change refers to proactive and reactive mechanisms by individuals, communities and nations to reduce their vulnerability and increase resilience to climate change impacts. As adaptive capacity is largely facilitated by access to resources, communities in poor and developing countries are considered to have low adaptive capacity because a large percentage of the population is often without access to the resources to withstand negative climate change effects. According to the 2001 IPCC assessment's chapter on Adaptation to Climate Change in the Context of Sustainable Development and Equity (Smit et al., 2001), policies that increase the adaptive capacity of communities, such as improved access to resources, and the lowering of inequality also help to achieve and maintain SDGs. Increased food insecurity, water scarcity and exposure to natural disasters undermine sustainable development objectives. Addressing those issues would inevitably play a critical role in climate change adaptation and mitigation efforts. Similarly, the 2007 IPCC report on Perspectives on Climate Change and Sustainable Development pointed to the impact of climate change on nations' ability to achieve SDGs. It reiterated that "the pace and character of development influences adaptive capacity and that adaptive capacity influences the pace and character of development" (Yohe et al., 2007).

Despite all that, climate change has often been treated as a separate environment policy block, without enough attention being paid to its social, economic and development linkages (Banuri & Opschoor, 2007; Halsnaes, Markandya & Shukla, 2011; Yohe et al., 2007). According to Banuri and Opschoor (2007), the conventional approach of separating climate change and economic development has actually led to increased climate change risks. Particularly with reference to developing countries, they argue that increasing emissions in pursuit of economic development has severely increased the risks of climate change in those countries. Greenhouse gas emissions from developing countries such as China and India have increased several hundred times over the last two decades, severely constraining their ability to demand more unilateral action by the developed countries. Those emissions, rarely subjected to strict national environment and health regulations have, in turn, negatively affected their own environmental standards and health status. For that reason, instead of pursuing fossil fuel-based energy production and consumption systems to foster economic growth, several scholars suggest a "decarbonisation" of the South (see Agarwal & Narain, 1999; Agarwal, Narain & Sharma, 2000) by increasing investments in renewable energy sources and building energy efficient products as a way to promote sustainable development and to adapt to climate change risks (also see Sathaye et al., 2007). In addition, developmental projects, such as poverty alleviation programs, often ignore the climate change dimension (Agrawala, 2004), thereby failing to attend to the increasing vulnerability and decreasing adaptive capacity of such groups, and thus missing several co-benefit opportunities.

Addressing the climate question would require an integrated set of measures and approaches that address other co-joint issues such as poverty reduction, economic development, increasing health status, among others – issues that impact vulnerability and adaptive capacity to climate change. Although some scholars from the Global South fear that climate change would displace the interest in and importance of other pressing issues such as poverty (Parikh, 1994), or would increase costs to address climate change specifically, there is a growing interest in policy making circles and emerging evidence to suggest that there are co-benefits in addressing climate change in alignment with development objectives, often termed as "mainstreaming" climate change policies (see TERI 2007; Metz et al., 2007).

Mainstreaming climate change into sustainable development

Mainstreaming of climate change policies refers to integrating climate change measures along with existing developmental policies to achieve the dual goals of sustainable development and climate change mitigation and adaptation (Halknes, Shukla, & Garg, 2008; Metz et al., 2007). Several studies indicate the potential and the benefits of mainstreaming climate change impacts and adaptation into development planning at global, national and local levels (e.g., Huq & Reid, 2004; Klein et al., 2007; Metz et al., 2007; TERI 2007; Tompkins, 2005). Developing countries, which are more likely to face the negative impacts of climate change, have a greater potential to integrate climate change into sustainable development objectives than developed countries because they are still in the process of infrastructure development. For example, although India's carbon emissions have increased by 80 per cent between 1990 and 2008 (IEA, 2011), over a third of Indian households still do not have access to electricity (India Census, 2011). Providing them with access to electricity is a critical development objective and would require India to build many new coal power plants – the predominant source of energy in India. That would achieve the development objective of increasing electricity access, but will result in a further increase in the levels of greenhouse gas emissions, thereby putting the Indian population at a higher risk for the effects of climate change, such as increased melting of the Himalayan glaciers, erratic monsoons and rising sea levels. Mainstreaming a climate change plan with the development objective of increasing energy access, therefore, would involve measures such as increasing energy efficiency and a shift to more renewable energy sources.

Such mainstreaming measures not only achieve the developmental goals and address climate change objectives, but can have other ancillary benefits such as decreasing local pollution, and thereby increasing local health and managing the increasing costs of oil imports (Sathaye et al., 2007). At the same time, other issues such as levels of corruption in a country could also affect the degree to which climate change is mainstreamed, and impact the achievement of developmental and climate change mitigation and adaptation goals (e.g., Mahmud & Prowse, 2012).

The response to mainstreaming climate change has been forthcoming, with several countries and other stakeholders seeking to mainstream climate change policies within the context of sustainable development. Some developing countries that will be the primary drivers of growth in this century have already shown leadership in sustainable development diplomacy through forming coalitions, such as Brazil, Russia, India, China and South Africa (BRICS) and Brazil, South Africa, India and China (BASIC), and identifying opportunities for joint cooperation and creating incentives for other key stakeholders to promote sustainability (Papa & Gleason, 2012). Similarly, large multilateral banks have begun to explicitly incorporate mitigation and adaptation objectives for funding development loans (World Bank, World Development Report 2010: Development and Climate Change). Moreover, an increasing number of multinational corporations, whose business may be affected as a result of climate change induced extreme weather events, have already started to explicitly address climate change plans as part of their larger organizational developmental objectives (e.g., UN Global Compact, http://www.unglobalcompact.org/). However, there are still gaps between the promise and implementation of mainstreaming climate change in sustainable development objectives and a need for a more urgent and ambitious shift by various stakeholders (e.g., Boyer, 2000; Lightfoot & Burchell, 2004; Sathaye et al., 2007; Srivastava, 2006; Plec & Pettenger, 2012).

Concluding remarks

Climate change is an urgent, and perhaps the greatest, threat that humanity faces in this century. Climate change vulnerability and adaptive capacity are inherently linked with the factors of poverty and access to resources and in many international forums they provide opportunities to focus on the larger imperative of sustainable development. Envisioning and implementing the SDGs, however, would require an *inclusive approach to the decision-making process*. The current policy decision process, negotiated in international contexts, is predominantly hierarchical, and the disenfranchisement of resources to vulnerable groups goes hand in hand with the marginalization of their power to communicate and be part of the policy making process (Dutta, 2012; Servaes, Polk, Shi, Reilly, & Yakupitijage, 2012). The erasure of the voices of indigenous groups, civil society and non-governmental organizations from international negotiations is incompatible with the basic principles of sustainable development that seek to foreground equity, sovereignty, and community participation. The transition to a sustainable development future would require respect to ingenious knowledge systems (Broadhead & Howard, 2011), common property rights and a more participatory approach (Servaes, 2013).

According to the World Bank's World Development Report 2010, addressing the climate change problem requires immediate action in all countries and a fundamental transformation of their energy systems including, among other things, a significant improvement in energy efficiency, a dramatic shift toward renewable energy and possibly nuclear power, and widespread use of advanced technologies to capture and store carbon emissions. Developed countries must lead the way and drastically cut their own emissions by as much as 80 per cent by 2050, bring new technologies to market, and help finance developing countries' transition to clean energy paths. But it is also in the developing countries' interests to act rapidly to avoid locking into a high-carbon infrastructure (World Bank, WDR 2010: Development and Climate Change). Moreover, developing countries' growth-first or growth-only objectives, without regard to climate change mitigation efforts, is facing increasing resistance from small island nations and the least developed countries (LDCs) – which are most threatened by climate change – which assert that all countries, including developing countries such as India, should commit to a strong and legally binding emission reduction plan.

The mainstreaming of climate change into sustainable development will continue to have a beneficial effect on the climate change talks.

There is obvious synergy in merging the two "frames" and a vast array of new opportunities to be explored by all stakeholders in pursuing the process further. Among them is the understanding and commitment of developed countries to sustainable development as a worthwhile effort in their own interest, which the focus on climate change mitigation and adaptation brings forth even more forcefully. From a development communication perspective, their convergence and addressing the new complex multidimensional issues involved may have been among the most exciting and challenging recent developments.

A great deal of work has been accomplished in the past 20 years, between the two Rio summits, and the authors are convinced that the work will be completed and the climate change negotiations will come to a successful end. It is unclear at this point whether this will be achieved in 2015, as some predict (Hedegaard, 2013), or (much) later. It is also hard to predict what will happen next: in other words, how much of climate change will be mainstreamed in sustainable development policies, how much and how fast decisions will be implemented, or how much political will and support such decisions will receive. One thing is certain, though: the best plans and intentions will only succeed with the full commitment of all institutions, all countries and all people, in a word – everyone.

Note

1. The opinions and views expressed in this chapter are made in the authors' personal capacity and should not be attributed to the institutions with which they are affiliated.

References

Agarwal, A. & Narain, S. (1999). *Addressing the Challenge of Climate Change: Equity, Sustainability and Economic Effectiveness: How Poor Nations Can Help Save the World.*

Agarwal, A., Narain, S., & Sharma, A. (2000). *Green Politics*. New Delhi, India: Center for Science and Environment.

Agrawala, S. (2004). Adaptation, development assistance and planning: Challenges and opportunities. *IDS Bulletin, 35*(3), 50–54. doi:10.1111/j.1759-5436.2004.tb00134.x.

Athanasiou,T. (1996). *The Age of Greenwashing. In Divided Planet: The Ecology of Rich and Poor*. Athens, GA: University of Georgia Press.

Banuri, T. & Opschoor, H. (2007). *Climate Change and Sustainable Development.* Desa Working Paper No. 56, ST/ESA/2007/DWP/56.

Boyer, B. (2000). Institutional mechanisms for sustainable development: A look at national councils for sustainable development in Asia. *Global Environmental Change, 10*(2), 157–160. doi:10.1016/S0959-3780(00)00019-4.

Broadhead, L.A. & Howard, S. (2011). Deepening the debate over "sustainable science": Indigenous perspectives as a guide on the journey. *Sustainable Development, 19*(5), 301–311. doi:10.1002/sd.421.

Brundtland, G. H. & World Commission on Environment and Development. (1987). *Our Common Future* (Vol. 383). Oxford: Oxford University Press.

Cohen, S., Demeritt, D., Robinson, J., & Rothman, D. (1998). Climate change and sustainable development: Towards dialogue. *Global Environmental Change, 8*(4), 341–371. doi:10.1016/S0959-3780(98)00017-X.

Dutta, M. (2012). *Voices of Resistance*. West Lafayette, IN: Purdue University Press.

Gawor, L. (2008). Globalization and its alternatives: Antiglobalism, alterglobalism and the idea of sustainable development. *Sustainable Development, 16*(2), 126–134. doi:10.1002/sd.336.

Gibson, R. (1991). Should environmentalists pursue sustainable development? *Probe Post*, 22–25.

Greer, J. & Kenny, B. (1996) *Greenwash: The Reality Behind Corporate Environmentalism*. Penang, Malaysia: Third World Network.

Halsnæs, K., Markandya, A., & Shukla, P. (2011). Introduction: Sustainable development, energy, and climate change. *World Development, 39*(6), 983–986. doi:10.1016/j.worlddev.2010.01.006.

Halsnæs, K., Shukla, P. R., & Garg, A. (2008). Sustainable development and climate change: Lessons from country studies. *Climate Policy, 8*(2), 202–219. doi:10.3763/cpol.2007.0475.

Hedegaard, C. (2013, 9 January). Focusing on 2015 climate deal. Retrieved from http://www.europolitics.info/sectoral-policies/focusing-on-2015-climate-deal-art346449-15.html

Huq, S. & Reid, H. (2004). Mainstreaming adaptation in development. *IDS Bulletin, 35*(3), 15–21. doi:10.1111/j.1759-5436.2004.tb00129.x.

IEA, International Energy Agency, 2011. *CO2 Emissions from Fuel Combustion 2010 – Highlights*. Retrieved from http://www.iea.org/co2highlights/co2highlights.pdf

India Census (2011). Houselisting and housing census data – 2011. Retrieved from http://censusindia.gov.in/

IPCC (2007). *Climate Change 2007: Synthesis Report. Contribution of Working Groups I, II and III to the Fourth Assessment Report of the Intergovernmental Panel on Climate Change* [Core Writing Team, Pachauri, R.K and Reisinger, A. (eds.)]. IPCC, Geneva, Switzerland, 104 pp.

IPCC (2012). *The IPCC Special Report on Managing the Risks of Extreme Events and Disasters to Advance Climate Change Adaptation*. Retrieved from http://www.ipcc.ch/news_and_events/docs/srex/SREX_slide_deck.pdf

Jayaraman, T. (2012). Holding back the "Green Economy" idea, but for how long? *Economic and Political Weekly, 47*(28). Retrieved from http://www.epw.in/commentary/holding-back-green-economy-idea-how-long.html

Jayaraman, T., Mittal, S., & Kohli, D. (2012). Rio + 20. *Economic and Political Weekly, 47*. Retrieved from http://www.epw.in/commentary/rio20.html

Khor, M. (2012). An assessment of the Rio summit on sustainable development. *Economic and Political Weekly, 47*(28). Retrieved from http://www.epw.in/commentary/assessment-rio-summit-sustainable-development.html

Klein, R. J. T., Eriksen, S. E. H., Næss, L. O., Hammill, A., Tanner, T. M., Robledo, C., & O'Brien, K. L. (2007). Portfolio screening to support the

mainstreaming of adaptation to climate change into development assistance. *Climatic Change, 84*(1), 23–44. doi:10.1007/s10584-007-9268-x.
Lightfoot, S. & Burchell, J. (2004). Green hope or greenwash? The actions of the European Union at the World Summit on sustainable development. *Global Environmental Change, 14*(4), 337–344. doi:10.1016/j.gloenvcha.2004.07.002.
Mahmud, T. & Prowse, M. (2012). Corruption in cyclone preparedness and relief efforts in coastal Bangladesh: Lessons for climate adaptation? *Global Environmental Change, 22*(4), 933–943. doi:10.1016/j.gloenvcha.2012.07.003.
Metz, B., Davidson, O.R., Bosch, P. R., Dave, R., & Meyer, L.A. (Eds.). (2007). *Climate Change 2007: Mitigation of Climate Change. Contribution of Working Group III to the Fourth Assessment Report of the Intergovernmental Panel on Climate Change*. Cambridge: Cambridge University Press.
Najam, A. (1999). World business council for sustainable development: The greening of business or a greenwash? In H.O. Bergesen, G. Parmann, & Ø.B. Thommessen (eds.,), *Yearbook of International Co-operation on Environment and Development 1999/2000* (pp. 65–75). London, UK: Earthscan Publications.
NOAA (n.d). NCDC Releases November 2012 U.S. Climate Report [Press release]. Retrieved from http://www.ncdc.noaa.gov/news/ncdc-releases-november-2012-us-climate-report
Papa, M. & Gleason, N. W. (2012). Major emerging powers in sustainable development diplomacy: Assessing their leadership potential. *Global Environmental Change, 22*(4), 915–924. doi:10.1016/j.gloenvcha.2012.06.003.
Parikh, J. (1994). North-South issues for climate change. *Economic and Political Weekly, 29*, 45–46. Retrieved from http://www.epw.in/special-articles/north-south-issues-climate-change.html
Plec, E., & Pettenger, M. (2012). Greenwashing consumption: The didactic framing of ExxonMobil's energy solutions. *Environmental Communication, 6*(4), 459–476.
Redclift, M. R. (1987). *Sustainable Development: Exploring the Contradictions*. Psychology Press.
Robinson, J. (2004). Squaring the circle? Some thoughts on the idea of sustainable development. *Ecological Economics, 48*(4), 369–384. doi:10.1016/j.ecolecon.2003.10.017.
Sanwal, M. (2011). Global vision for Rio + 20 and beyond. *Economic and Political Weekly, 46*(40). Retrieved from http://www.epw.in/commentary/global-vision-rio20-and-beyond.html
Sathaye, J., Najam, A., Cocklin, C., Heller, T., Lecocq, F., Llanes-Regueiro, J., Pan, J., Petschel-Held, G., Rayner, S., Robinson, J., Schaeffer, R., Sokona, Y., Swart, R., & Winkler, H. (2007), Sustainable development and mitigation. In B. Metz, O.R. Davidson, P.R. Bosch, R. Dave, & L.A. Meyer (Eds.) *Climate Change 2007: Mitigation. Contribution of Working Group III to the Fourth Assessment Report of the Intergovernmental Panel on Climate Change*. Cambridge, United Kingdom and New York, NY, USA: Cambridge University Press.
Servaes, J., Polk, E., Shi, S., Reilly, D., & Yakupitijage, T. (2012). Towards a framework of sustainability indicators for "communication for development and social change" projects. *International Communication Gazette, 74*(2), 99–123. doi:10.1177/1748048511432598.
Servaes, J. (ed.) (2013). *Sustainability, Participation and Culture in Communication. Theory and Praxis*, Bristol/Chicago: Intellect/University of Chicago Press.

Smit, B., Pilifosova, O., Burton, I., Challenger, B. Huq, S., Klein, R.J.T., & Yohe, G. (2001). Adaptation to climate change in the context of sustainable development and equity. Chapter 18 in *Climate Change 2001: Impacts, Adaptation, and Vulnerability – Contribution of Working Group II to the Third Assessment Report of the Intergovernmental Panel on Climate Change*. Cambridge, UK: Cambridge University Press.

Srivastava, L. (2006). Climate protection for sustainable development or sustainable development for climate protection? A case study from India. *Global Environmental Change, 16*(2), 120–122. doi:10.1016/j.gloenvcha.2006.01.005.

Stern, N. (2007). *Economics of Climate Change: The Stern Review*. Cambridge, UK: Cambridge University Press.

TERI (2007) *Adaptation to Climate Change in the Context of Sustainable Development*. New Delhi, India: The Energy and Resources Institute. Retrieved from www.teriin.org/events/docs/adapt.pdf

Tokar, B. (1997). *Earth for Sale: Reclaiming Ecology in the Age of Corporate Greenwash*. Boston, MA: South End Press.

Tompkins, E. L. (2005). Planning for climate change in small islands: Insights from national hurricane preparedness in the Cayman Islands. *Global Environmental Change, 15*(2), 139–149. doi:10.1016/j.gloenvcha.2004.11.002.

U.N. (1987). Report of the World Commission on Environment and Development. General Assembly Resolution 42/187, 11 December 1987.

UNDP (2007–08). Fighting climate change: Human solidarity in a divided world. Retrieved from http://hdr.undp.org/en/media/HDR_20072008_EN_Complete.pdf

Vidal, J. (2012, 27 June). Rio + 20: Reasons to be cheerful. *The Guardian*. Retrieved from http://www.guardian.co.uk/global-development/poverty-matters/2012/jun/27/rio20-reasons-cheerful

WMO (2012). WMO provisional statement on the state of global climate in 2012 [Press release]. Retrieved from http://www.wmo.int/pages/mediacentre/press_releases/documents/966_WMOstatement.pdf

World Bank. (2010). *WDR 2010: Development and Climate Change*. Washington DC, USA: World Bank.

Yohe, G.W., Lasco, R.D., Ahmad, Q.K., Arnell, N.W., Cohen, S.J., Hope, C., Janetos, A.C., & Perez, R.T. (2007). Perspectives on climate change and sustainability. In M.L. Parry, O.F. Canziani, J.P. Palutikof, P.J. van der Linden & C.E. Hanson (Eds.), *Climate Change 2007: Impacts, Adaptation and Vulnerability. Contribution of Working Group II to the Fourth Assessment Report of the Intergovernmental Panel on Climate Change* (pp. 811–841). Cambridge, UK: Cambridge University Press. Retrieved from www.ipcc.ch/pdf/assessment-report/ar4/wg2/ar4-wg2-chapter20.pdf

2

Future Imperatives of Practice: The Challenges of Climate Change

Chin Saik Yoon

Upheavals in practice

Practitioners of communication for development and social change are soon to be severely tested by the "new" issues that will preoccupy their work in the future. The new issues will challenge practitioners as they pose systemic questions for which answers are not readily available. These escalating imperatives of development include (but are not limited to):

- climate change;
- financial and economic crisis;
- competition for natural resources;
- political transitions in young democracies; and
- warfare and terrorism.

Embedded in these challenges are specific issues related to our ecosystems, loss of land and territories, food security, health, employment, gender, culture, oppression of minority groups, conflicts and migration. Most of these issues have no ready solutions for communication and action and are without precedence. They will test the ingenuity of communicators now and in the generations to come.

Unlike past development issues arising from food and agriculture, public health, education and income generation, the new issues are all-encompassing. Communication practitioners are just beginning to understand the salient aspects of what some of the issues involve, and how they impact on communities. This educational process of practitioners will need to continue while we begin to evolve processes of practice to begin responding to the challenges they pose. At the same time we will need to retool our skill-sets to prepare ourselves for work on the ground and engagement with the communities to which we belong. This chapter will try to outline what it is we do not know about the emerging imperatives so that we can better appreciate the challenges we face.

This chapter will refer only to the challenges posed by climate change – one of the five large clusters of "new" issues – so as to keep the discussions focused and manageable. However, because climate is central to the well-being of the planet, it is hoped that the questions this chapter is raising will also contribute toward the opening of our discussions about the numerous development issues of the future.

A brief overview of climate change

The politics behind climate change have led to sharply opposing positions among experts. It has even distorted scientific evidence. The stakes are high and many have lots to gain if scientific data showing evidence of climate change are successfully denied and the status quo maintained. Much of the denial originated from within industry while the majority of advocates are members of civil society. This polarization was clearly evident in the United States. Former Vice President Al Gore presented comprehensive proof of global warming in his famous Powerpoint slide show, *An Inconvenient Truth.*[1] Opposing Gore's analysis was the United States government, then led by President Bush, who downplayed the evidence and countered with their own significantly more positive scenario. James Hansen, Director of NASA's top research center on climate change was asked in 2006, on the television program *60 Minutes,* whether he was being censored by the Bush Administration about what he could say in public about global warming; he famously replied: "Or they're censoring whether or not I can say it. I mean, I say what I believe if I'm allowed to say it."[2] The need to manoeuvre around these sharp differences in positions has become, by itself, a challenge to practitioners.

Al Gore's evidence of climate change includes a fairly extensive synthesis of research showing the linkage between continuous increases in the level of greenhouse gases (GHG) in the atmosphere to increases in the earth's temperatures. He referred to work done by scientists examining ice cores that date back over 650,000 years to support this conclusion. Gore also examined time-series remote sensing images that show the retreat of glaciers around the world. The shrinkage of these glaciers is deeply worrying as they are the sources of many rivers that provide populous regions with their supply of fresh water. The accelerated melting of these glaciers is ominous as it signals the impending reduction of the flow and the eventual drying up of rivers in generations to come. Many of the rivers affected are found in the Asia-Pacific region.

Impact of climate change on agriculture

Changes in the climate will become quickly apparent in agriculture. Variations in rainfall, reduced river flow, salinization of agricultural land and changes in temperature will affect our ability to grow sufficient amounts of food for the population. Experts at the Food and Agriculture Organization (FAO), part of the United Nations, presented the following scenario in its 2009 publication *Profile for Climate Change*[3]:

> **Selected projected regional impacts of climate change on food production**
>
> The higher latitudes might experience productivity increases due to higher temperatures, longer vegetation periods and CO_2 enrichment. However, in tropical and sub-tropical regions, especially in seasonally dry areas, crop and animal productivity may decrease significantly with temperature increases of 2–3°C causing for example:

- Significant decreases in crop yields in some rain-fed African systems,
- Declines of 40–80 per cent of grassland productivity in semi-arid and arid regions,
- High levels of desertification and soil salinization in some areas in Asia, sub-Sahara Africa and Latin America, and
- Increasing water stress, particularly in irrigated production systems.

Areas such as sub-Saharan Africa and South Asia may benefit from increased annual rainfall but some of these areas, such as coastal plains, may become flooded or unsuitable for agriculture due to increased salinity

> In the fisheries sector, ecosystem productivity is likely to be reduced in most tropical and subtropical oceans, seas and lakes due to habitat displacement towards higher latitudes. However, climate change also provides new opportunities for aquaculture, particularly in Asia.

The scenario above reveals another challenge for communicators: while most people will suffer unfavourable consequences of climate change, some of us may temporarily enjoy the opposite and paradoxically gain from the "positive" effects of climate change. These paradoxes are likely to confuse communication and mobilization efforts, and may be used by groups working to stall or neutralize policies aimed at mitigating the causes of climate change.

Africa is a region that will bear the brunt of some of the most serious effects of climate change. FAO's 26th Regional Conference for Africa, held in Angola in May 2010, noted[4] that the effects are already apparent; a marked decline in rainfall has already led to a 40–60 per cent discharge in rivers in parts of the region. This in turn has reduced the availability of surface water. The reduction is clearly evident in the shrinkage of the area covered by Lake Chad from 20,000 sq km prior to the 1970s to its current 7,000 sq km.

The reduction of water supply will impact most aspects of agriculture. Less drought-resistant crops will fail. Pastoralists will find it increasingly difficult to provide sufficient water for their herds, while at the same time areas available for grazing will shrink. The impact will be felt not only on land but also in water: fishing-dependent communities will face a reduced stock of fish to catch in depleted rivers. Climate change will undermine food security and the poor will be affected disproportionately.

Impact of climate change on health

The World Health Organization (WHO)[5] anticipates that changes in climatic conditions will lead to three kinds of health impacts:

- Those that are relatively direct, usually caused by weather extremes.
- The health consequences of various processes of environmental change and ecological disruption that occur in response to climate change.
- The diverse health consequences – traumatic, infectious, nutritional, psychological and other – that occur in demoralized and displaced populations in the wake of climate-induced economic dislocation, environmental decline, and conflict situations.

WHO assessed some of the health outcomes that are sensitive to climate variation. Four types of outcomes were identified[6]:

- Food and water-borne disease – leading to diarrhoea episodes
- Vector-borne disease – leading to malaria cases
- Natural disasters involving coastal and inland floods and landslides – leading to fatal injuries

- Risk of malnutrition – causing non-availability of recommended daily calorie intake.

Assessment of the Intergovernmental Panel on Climate Change

Highlights of the above FAO and WHO assessments provide us with just glimpses of a tiny sample of the wide range of issues connected to climate change. The World Meteorological Organization (WMO) and the United Nations Environment Program (UNEP) set up The Intergovernmental Panel on Climate Change (IPCC) in 1988 to provide the governments of the world with a clear scientific view of what is happening to the world's climate. The United Nations (UN) General Assembly approved Resolution 43/53 in December of the same year, which provides the IPCC with the mandate to prepare a comprehensive review and make recommendations on climate change. The first assessment of the salient aspects of climate change – including: current scientific knowledge, the social and economic impact anticipated, possible response strategies and elements for an international convention on climate – led to the publication of the first report on climate change in 1990. Continuing work by the panel over the following two decades led to the publication of three additional reports in 1995, 2001 and 2007.[7] These reports provide us with a comprehensive and authoritative picture of the challenges we face.

The succinct *Climate Change 2007: Synthesis Report*[8] is a distillation of the findings of the most recent assessment of the IPCC. It aims to provide policy makers with a concise and more accessible overview of the scientific opinion of the thousands of scientists involved in the fourth round of assessment. Communicators will also find this synthesis useful reading. A figure and two tables from IPCC's report are reproduced below as a primer for our discussions about communication for development and social change in relation to climate change.

Figure 2.1 is from page 10 of the *Synthesis Report*. It describes the impacts we can expect in the event of global warming of five degrees Celsius. The figure looks at impact on five fronts: water for human use, ecosystems, food, coastal land areas and human health. More importantly, it shows we are on the brink of the tipping point that will see us sink into a very uncomfortable climate regime which is likely to beset us with droughts, floods, extinction of species and a reduced supply of food.

I have next extracted from Table SPM.2 (Figure 2.2), found on page 11 of IPCC's *Synthesis Report,* which includes the forecast for Africa – the region that is set to become one of the worst affected. Also included are

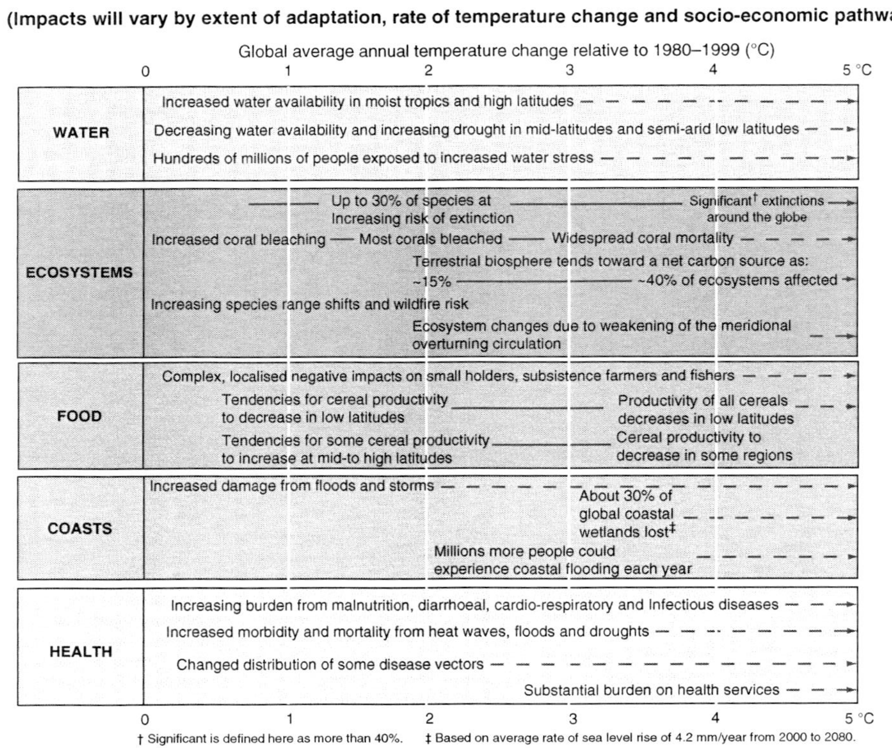

Figure 2.1 Examples of impacts associated with global average temperature change (IPCC)

Africa	• By 2020, between 75 and 250 million of people are projected to be exposed to increased water stress due to climate change. • By 2020, in some countries, yields from rain-fed agriculture could be reduced by up to 50%. Agricultural production, including access to food, in many African countries is projected to be severely compromised. This would further adversely affect food security and exacerbate malnutrition. • Toward the end of the 21st century, projected sea level rise will affect low-lying coastal areas with large populations. The cost of adaptation could amount to at least 5 to 10% of Gross Domestic Product (GDP). • By 2080, an increase of 5 to 8% of arid and semi-arid land in Africa is projected under a range of climate scenarios (TS).
Asia	• By the 2050s, freshwater availability in Central, South, East and South-East Asia, particularly in large river basins, is projected to decrease. • Coastal areas, especially heavily populated megadelta regions in South, East and South-East Asia, will be at greatest risk due to increased flooding from the sea and, in some megadeltas, flooding from the rivers. • Climate change is projected to compound the pressures on natural resources and the environment associated with rapid urbanization, industrialization and economic development. • Endemic morbidity and mortality due to diarrhoeal disease primarily associated with floods and droughts are expected to rise in East, South and South-East Asia due to projected changes in the hydrological cycle.
Australia and New Zealand	• By 2020, significant loss of biodiversity is projected to occur in some ecologically rich sites, including the Great Barrier Reef and Queensland Wet Tropics. • By 2030, water security problems are projected to intensify in southern and eastern Australia and, in New Zealand, in Northland and some eastern regions. • By 2030, production from agriculture and forestry is projected to decline over much of southern and eastern Australia, and over parts of eastern New Zealand, due to increased drought and fire. However, in New Zealand, initial benefits are projected in some other regions. • By 2050, ongoing coastal development and population growth in some areas of Australia and New Zealand are projected to exacerbate risks from sea level rise and increases in the severity and frequency of storms and coastal flooding.
Small Islands	• Sea level rise is expected to exacerbate inundation, storm surge, erosion and other coastal hazards, thus threatening vital infrastructure, settlements and facilities that support the livelihood of island communities. • Deterioration in coastal conditions, for example through erosion of beaches and coral bleaching, is expected to affect local resources. • By mid-century, climate change is expected to reduce water resources in many small islands, e.g. in the Caribbean and Pacific, to the point where they become insufficient to meet demand during low-rainfall periods. • With higher temperatures, increased invasion by non-native species is expected to occur, particularly on mid- and high-latitude islands.

Figure 2.2 Examples of some projected regional impacts (IPCC)

those cells of the table that look ahead to the future of the Asia-Pacific region. The forecast is that we are going to be hit by an ironical mix of droughts and floods.

Bangkok, for example, which is one of the largest Asian cities, is forecasted to suffer from worse than average flooding due to the rise of sea levels exacerbated by increased flows from the river and waterways that surround the city.

In the case of the Pacific Islands, we will witness an increasing percentage of land becoming submerged under the rising sea, accompanied by significant reduction in local fresh water resources that will undermine the sustainability of the population during the dry seasons.

Table SPM.3 (Figure 2.3) from page 13 of IPCC's *Synthesis Report* takes a look at where the current trend of warming will take us by the mid-twenty-first century; the younger readers of this publication will most likely live to experience the forecasted consequences.

Phenomenon[a] and direction of trend	Likelihood of future trends based on projections for 21st century using SRES scenarios	Examples of major projected impacts by sector			
		Agriculture, forestry and ecosystems	Water resources	Human health	Industry, settlement and society
Over most land areas, warmer and fewer cold days and nights, warmer and more frequent hot days and nights	*Virtually certain*[b]	Increased yields in colder environments; decreased yields in warmer environments; increased insect outbreaks	Effects on water resources relying on snowmelt; effects on some water supplies	Reduced human mortality from decreased cold exposure	Reduced energy demand for heating; increased demand for cooling; declining air quality in cities; reduced disruption to transport due to snow, ice; effects on winter tourism
Warm spells/heat waves. Frequency increases over most land areas	*Very likely*	Reduced yields in warmer regions due to heat stress; increased danger of wildfire	Increased water demand; water quality problems, e.g. algal blooms	Increased risk of heat-related mortality, especially for the elderly, chronically sick, very young and socially isolated	Reduction in quality of life for people in warm areas without appropriate housing; impacts on the elderly, very young and poor
Heavy precipitation events. Frequency increases over most areas	*Very likely*	Damage to crops; soil erosion, inability to cultivate land due to waterlogging of soils	Adverse effects on quality of surface and groundwater; contamination of water supply; water scarcity may be relieved	Increased risk of deaths, injuries and infectious, respiratory and skin diseases	Disruption of settlements, commerce, transport and societies due to flooding; pressures on urban and rural infrastructures; loss of property
Area affected by drought increases	*Likely*	Land degradation; lower yields/crop damage and failure; increased livestock deaths; increased risk of wildfire	More widespread water stress	Increased risk of food and water shortage; increased risk of malnutrition; increased risk of water- and food-borne diseases	Water shortage for settlements, industry and societies; reduced hydropower generation potentials; potential for population migration
Intense tropical cyclone activity increases	*Likely*	Damage to crops; windthrow (uprooting) of trees; damage to coral reefs	Power outages causing disruption of public water supply	Increased risk of deaths, injuries, water- and food-borne diseases; post-traumatic stress disorders	Disruption by flood and high winds; withdrawal of risk coverage in vulnerable areas by private insurers; potential for population migrations; loss of property
Increased incidence of extreme high sea level (excludes tsunamis)[c]	*Likely*[d]	Salinisation of irrigation water, estuaries and fresh-water systems	Decreased fresh-water availability due to saltwater intrusion	Increased risk of deaths and injuries by drowning in floods; migration-related health effects	Costs of coastal protection versus costs of land-use relocation; potential for movement of populations and infrastructure; also see tropical cyclones above

An eclectic mix of impacts is expected in about 50 years. There will be increasingly severe droughts and storms, all of which will badly affect communities. The sea level may rise and introduce salt into the soil of

Key PDC processes

Effective communication

Self expression
Listening
Understanding

Creating knowledge

Sourcing information
Tapping indigenous knowledge
Processing and validating information
Sharing knowledge

Building communities

Building trust
Managing conflict and competition
Forging partnerships
Reinforcing self identity
Reflecting on the past and present
Visioning the future
Affirming values
Adjusting values
Enabling transparency in decision making
Sharing benefits

Enabling action

Identifying problems
Evolving solutions
Nurturing sense of guardianship of the commons
Managing expectations
Taking stock and pooling resources
Sourcing complimentary resources
Advocating to stakeholders
Mobilizing for action
Evaluating action
Iterating and refining action

Figure 2.3 Key PDC processes

low-lying land areas. Sea water is also forecasted to extend upstream of waterways and taint river estuaries and irrigation systems. The fresh water in these waterways will be rendered not suitable for human activities when this occurs.

The new imperatives

How is working on climate change going to be different from past development communication projects?

I think it will be different on four important fronts: goals, scale, time and altruistic action.

Some of these "differences" have been emerging from our past and current work on "macro" issues related to sustainable development, natural resources management, farming systems research, human rights, values and religions, HIV/AIDS and culture.

Goal

Past development communication projects have succeeded because communities were motivated by tangible benefits of the goals of projects we were involved in. Farmers were able to see increased yields at the end of a growing season; mothers could see their babies grow healthier from week to week; and backyard entrepreneurs saw the sale of their products increase. Climate change projects promise none of that. Climate change initiatives will seek to do the opposite – promoting less consumption of natural resources by everyone so that "the commons", belonging to no one, can heal. Solutions to global warming will often contradict past development communication messages. These solutions will offer none of the short- and medium-term benefits that past development projects promised.

Some governments have tried to frame climate change action in the existing rubric of industry by saying that manufacturers and workers can derive substantial economic profits from the production of new "green technologies" powered by alternative sources of energy. While this may be true, it fails to address the equally urgent problems of water pollution and loss of forests – both these problems can be made worse in the course of manufacturing even the new green technologies, just as has happened with the old technologies.

Scale

Communicators must situate and orientate their efforts within a very large context even if they elect to work locally. As the climate is derived from a region- or planet-wide dynamic, the events happening thousands of miles away can have a profound impact at our doorstep. As such, communicators – and the communities they serve – may find themselves unable to address the causes of their development problems brought about by climate change even if they are mobilized to act in a concerted way to address the root causes.

Time

Unlike in the past when the benefits of social change and development became evident in a cropping cycle, or within a few years of the launching of an enterprise, our efforts to address climate change may not become evident for a couple of generations – long after present-day initiators of change have passed away from the scene. How does one initiate projects planned for continuous work across generations? How do the generations of communicators sustain such efforts without any evidence or promise of benefit within their lifetimes? In what ways should development funding agencies re-orientate their programs and modalities so that their current tranches of two- to five-year grants can help to initiate projects that should carry on over to the next generation for them to make real impact on the mitigation of climate change?

Altruistic action

Central to the success of climate change projects will be the revival of the concept of "the commons" and the suppression of individualistic interests, which most development projects in the past promoted. Mitigation of the causes of climate change can only be achieved if we succeed in tapping into the altruistic values and principles of communities. People must be willing to lead simpler lives so that their descendents – whom they will never meet – stand a better chance of survival. This involves communities taking "social responsibility" for present and future generations. It will involve making sacrifices here and now so that people in the most affected regions may better adapt to the harsher climate regimes elsewhere; and help ensure that the generations not yet born may be spared the harsher impacts that are yet to come.

Two-prong response to climate change

Experts in climate change recommend two main sets of responses to climate change: "adaptation" to the effects of climate change; and the "mitigation" of its causes – currently focused on the reduction of GHGs in the atmosphere. GHGs comprise (listed here in their order of abundance): water vapour, carbon dioxide, methane, nitrous oxide, ozone and chlorofluorocarbons.

The United Nations Framework Convention on Climate Change[9] synthesized the following points regarding possible response for us to consider.

Adaptation options:

- Adaptation can reduce vulnerability, both in the short and the long term;
- Vulnerability to climate change can be exacerbated by the presence of other stresses, arising for example from current climate

hazards, poverty, unequal access to resources, food insecurity, trends in economic globalization, conflict and incidence of diseases such as HIV/AIDS;

- Adaptation will be required at regional and local levels to reduce the adverse impacts of projected climate change and variability, regardless of the scale of mitigation undertaken;
- A wide array of adaptation options is available, but more extensive adaptation than is currently occurring is required to reduce vulnerability to future climate change. There are barriers, limits and costs, but these are not fully understood;
- Adaptive capacity is intimately connected to social and economic development, but it is not evenly distributed across and within societies. Mitigation options:
- There is substantial economic potential for the mitigation of global GHG emissions over the coming decades, that could offset the projected growth of global emissions or reduce emissions below current levels (high agreement and much evidence);
- In all analyzed world regions, near-term health co-benefits from reduced air pollution, as a result of actions to reduce GHG emissions, can be substantial and may offset a substantial fraction of mitigation costs (high agreement and much evidence);
- There may be effects from Annex I countries [developed countries and economies in transition] action on the global economy and global emissions, although the scale of carbon leakage remains uncertain (high agreement, medium evidence);
- Changes in life style and behaviour patterns can contribute to climate change mitigation across all sectors. Management practices can also have a positive role (high agreement, medium evidence);
- Policies that provide a real or implicit price of carbon could create incentives for producers and consumers to significantly invest in low-GHG products, technologies and processes; and
- A wide variety of national policies and instruments are available to governments to create the incentives for mitigation action (high agreement and much evidence). Their applicability depends on national circumstances and an understanding of their interactions, but experience form implementation in various countries and sectors shows there are advantages and disadvantages for any given instrument.

Practitioners of communication have meaningful roles to play in both prongs of response. However, the more daunting challenges at the local level lie in the "adaptation" to climate change. This is probably the "frontline", where most practitioners of communication will be deployed. We will face there the real-life, day-to-day struggles of people impacted by droughts, storms, submerged farms and homes, diseases, depleted supplies of potable water and shortage of affordable food. That is where many of the new imperatives will be the most challenging.

Much of the communicators' work on the "mitigation" responses are now focused on the support of advocacy work by various groups engaged with the troubled "climate talks" at the UN. Our colleagues in the media have played a central role in raising public awareness about climate change. Campaigners leading "climate groups" in civil society have also been effective at raising awareness about the issue; more importantly, they have been successful in bringing political pressure to bear on elected officials to temper their response to the powerful lobbies of industries and to heed the concerns of their electorate. They are building and enlarging the constituencies that policy makers engaged in the multilateral negotiations must answer to. They have also in the past faced off against the industrial lobby and their attempts to maintain the status quo and minimize the extent of legislation and conventions to limit GHG emissions.

When the UN does eventually succeed in approving effective and binding mitigation measures that member states must comply with, the work of both the media and citizen groups in mitigation will most likely shift to a combination of "watch-dog" and information provider. The watch-dog role will probably be played by both the mass media and social media to monitor compliance by governments and industry; and at the same time blow the whistle on those that cheat. The information provider role may involve the media sharing across borders, industry and professions, experiences and expertise on mitigation methods to help GHG emitters with their efforts to comply. Over time, technologies may emerge where GHGs existing in the atmosphere may be extracted from the air and safely sunk back into the earth; it will also be the role of communicators to help disseminate information about these new technologies as well.

Clues for future practice

Past and current development communication training has not fully equipped practitioners with the necessary skills to work on climate change issues. Information and experiences about the missing elements of practice may come from eroded tenets of our cultures, indigenous knowledge, traditional societies and organized religions – the majority of which engender sustainable living and respect for nature and the intrinsic.

Adopting an inter-disciplinary approach

We should also keep looking beyond the traditional boundaries of communication and learn from the allied disciplines and sciences. Psychology is one such science we ought to tap. The Australian Psychological Society published a position chapter on climate change in August 2010, where it identified the following issues[10] as central to psychology and climate change; you will also recognize these as being equally critical to our work in communication for social change:

> **6. People's risk appraisals and understanding of climate change**
>
> 6.1 How people perceive, appraise and understand environmental problems is critical because these processes affect how concerned people are (or not), and how motivated they are to take action (or not).
>
> 6.2 Psychologists are uniquely positioned to explore how people's emotional reactions and other moderating factors combine with rational thought to produce a sense of risk and the willingness to take actions that reduce it.
>
> 6.3 Understanding why scientists and other relevant stakeholders may not get their messages about risk across to the public – and why they are not always able to motivate appropriate action even when they do – is one of the areas where psychology can help address the challenge of climate change.
>
> **7. Psychological contributions to how people change behaviour**
>
> 7.1 Human behaviour is complex, and deeply embedded in social situations, institutional contexts and cultural norms. Psychologists contribute valuable knowledge about the importance of addressing numerous and often interacting factors including risk perceptions and understandings, threat and coping response appraisals, the extent to which individuals think that their actions will make a difference, barriers to action, and ways of motivating people to change.
>
> 7.2 Barriers to action include structural barriers (e.g., lack of public transport), socio-cultural barriers such as perception of high status

coming from large cars and large homes, and economic disincentives (e.g., high costs of "greening" behaviours, especially for low income families). There are also significant psychological barriers including ignorance, uncertainty, mistrust, denial, judgmental discounting, lifetime habits, perceived risks, tokenism, social comparison, conformity, and conflicting goals and aspiration.

7.3 Psychologists have a long history of work in motivating change.

7.4 Motivating change includes a range of strategies that have been identified in the social science literature for helping people to overcome barriers. People need to know what they can do, what actions they can take, and why taking such actions can really make a difference. Both the public and policy makers need to be aware of the relative efficacy and utility of particular actions. Psychologists can identify and promote strategies which directly address people's concerns, anxieties and distress in the face of a daunting set of environmental issues and uncertainties, and can also identify environmentally significant behaviours with an emphasis on psychologically significant responses that lead to feelings of efficacy, responsibility, hope and optimism.

7.5 As well as having expertise in motivating and facilitating change at an individual level (including intra-individual variables as well as behaviours), psychologists also have expertise in developing an understanding of broader physical, geophysical and social environments, including infrastructure, landscaping, weather, scarcity of resources, sense of community and feeling of "ownership" of a place. They can also advise on how to measure and modify aspects of these environments to achieve sustainable change.

7.6 Both bottom-up (e.g., individual and community commitment), and top-down (e.g., government leadership and legislative strategies) initiatives are needed. Behavioural science knowledge is critical at all levels of the design, implementation and evaluation of proenvironmental campaigns and activities at an individual and societal level. In addition, social scientists can develop models of decision making that will help policy-makers understand how the public will respond once policies are put into place, and advise government on how to ensure that climate policies are effectively communicated and turned into effective actions.

Adopting a participatory approach

Within the discipline of communication, our past work on participatory communication offers the most promising concepts, strategies and methodologies for the challenges ahead. Climate is the key component of our commons. Care for the commons cannot be successfully undertaken in a fragmented manner. Nor can it be imposed upon the planet in a top-down manner. It can only be accomplished if everyone participates in the healing of our commons.

We should build on the seminal work of colleagues such as Jan Servaes[11] and particularly his multiplicity paradigm. We should also learn from past innovations in practice, which have succeeded. A number of our fellow practitioners have valuable experiences to share.

Tim Kennedy, who started off as a novice communications practitioner – and is now a senior professor of the discipline – undertook one of the earliest experiments in participatory communication. The social impact of his early work is still clearly apparent today. He is a pioneer with a track record to inspire. His work among the native people of Alaska living in the extreme climatic conditions of the Arctic more than 40 years ago is highly relevant to the climate change issues of today. Tim developed the participatory film-making method called The SKYRIVER Process, to mobilize the native people to speak out against the ill-informed and badly designed development projects imposed on them by American officials from outside Alaska. In the process he strengthened the community, affirmed their culture, and focused the people's desire and power to bring about change. He recently returned to his project site after an absence of 40 years to discover the long-lasting and extensive impact of the the SKYRIVER Process. He heard moving personal stories of the lives touched and altered for the better by his work as a practitioner four decades ago. His experiences and the lessons which he learnt from his ground-breaking work have recently been published in his book *Where The Rivers Meet The Sky.*[12]

The more recent work of Guy Bessette and the participatory development communication (PDC) model[13] of action and research is also highly relevant. The PDC model evolved out of a network of researchers working on natural resource management issues. The network included active research groups from Africa, Asia and the Arab Region. The problems that the network members addressed touch on most of the adaptation options for climate change. The experiences of the network were synthesized in the book *People, Land and Water: Participatory development communication for natural resource management*[14] edited by Bessette.

The author was provided with the opportunity to be a student of all the research teams and contributed the summary chapter to the book.[15] The consolidated work of the PDC network showed the progress that both practitioners and researchers had made in adopting participatory approaches. The experiments also revealed the many processes that we still do not know how to work with in practice. These processes are all relevant and important to issues relating to climate change.

The processes that PDC practitioners concentrate on facilitating may be clustered into four categories:

- effective communication;
- creating knowledge;
- building communities; and
- enabling action.

The processes which make up each of these categories are listed in the figure below.

The availability and degree of absence of "tools", methods and strategies for facilitating the processes listed in the figure above are summarized in the next matrix.

The imperatives

The matrix above shows that many of the missing or emerging tools and methods are clustered around the two areas of creating knowledge and building communities.

These PDC experiments show that many of the participatory processes are interpersonal in nature; or involve participatory group processes that seem best served by interpersonal or group communication techniques. This indicates the need for a range of interpersonal communication techniques and group facilitation methods to be developed or acquired by practitioners. People-embodied methods are some of the most challenging to develop and to share. An effective facilitator of participatory group processes is easy to appreciate but very difficult to emulate.

The most challenging tool that practitioners must acquire for tackling climate change is the skill of managing conflicts. Communities and individuals engaged in adaptation to the impact of climate change, or advocacy for the mitigation of its causes, frequently find themselves in conflict with the positions of other groups, and in competition for declining resources that others need as well.

PROCESSES	AVAILABILITY OF TOOLS AND METHODS		
	Exists	***Being Developed***	***Largely Absent***
Effective Communication			
Self expression	✓		
Listening and understanding		✓	
Creating Knowledge			
Sourcing information	✓		
Tapping indigenous knowledge		✓	
Processing and validating information		✓	
Sharing knowledge			✓
Building Communities			
Building trust		✓	
Managing conflict and competition			✓
Forging partnerships		✓	
Reinforcing self identity			✓
Reflecting on the past and present	✓		
Visioning the future	✓		
Affirming values			✓
Adjusting values			✓
Enabling transparency in decision making			✓
Sharing benefits		✓	
Enabling Action			
Identifying problems	✓		
Evolving solutions	✓		
Nurturing sense of guardianship of the commons		✓	
Managing expectations		✓	
Taking stock and pooling resources	✓		
Sourcing complimentary resources	✓		
Advocating to stakeholders	✓		
Mobilising for action	✓		
Evaluating action	✓		
Iterating and refining action	✓		

Figure 2.4 Availability of tools and methods for PDC processes

These skills and tools are not only vital for tackling climate change but they will also come in very handy in addressing the other exigencies in development we are facing or may soon have to face:

- financial crisis;
- competition for natural resources;
- political transitions in young democracies; and
- warfare and terrorism.

Notes

1. Gore, Al (2006) *An Inconvenient Truth*, Hollywood: Paramount Pictures, information about the DVD available for download at http://www.climatecrisis.net/an_inconvenient_truth/about_the_film.php
2. *60 Minutes* CBS (2006) Rewriting the science: Scientist says politicians edit global warming research, available for download at http://www.cbsnews.com/stories/2006/03/17/60minutes/main1415985.shtml
3. FAO (2009) *Profile for Climate Change*, Rome: FAO, p. 3.
4. FAO (2010) "Climate Change Implications for Food Security and Natural Resources Management in Africa", *Twenty-Sixth Regional Conference for Africa*, Launda, Angola 03–07 May 2010, Document ARC/10/8 available for download at http://www.fao.org/docrep/meeting/018/k7542e.pdf.
5. WHO (2003) Climate change and human health – risks and responses. Summary, available for download at http://www.who.int/globalchange/summary/en/index2.html
6. WHO (2003) How much disease would climate change cause? available for download at http://www.who.int/globalchange/summary/en/index6.html
7. IPCC (1990, 1995, 2001, 2007) Assessment Report: Climate Change, PDF and electronic versions of these reports are available for download at: http://www.ipcc.ch/publications_and_data/publications_and_data_reports.shtml
8. IPCC, *Climate Change 2007: Synthesis Report*, Geneva: IPCC available for download at http://www.ipcc.ch/publications_and_data/ar4/syr/en/spm.html
9. United Nations Framework Convention on Climate Change (2010) Fact sheet: Climate change science – the status of climate change science today, available for download at http://unfccc.int/press/fact_sheets/items/4987txt.php.
10. The Australian Psychological Society (2010) *Psychology and Climate Change: A Position Statement Prepared for The Australian Psychological Society*, pp. 6–7, available for download at http://www.psychology.org.au/Assets/Files/Position-statement-climate-change-August2010.pdf
11. Servaes, Jan (1999) *Communication for Development; One World, Multiple Cultures*. Cresskill: Hampton Press.
12. Kennedy, Timothy (2008) *Where The Rivers Meet The Sky: A Collaborative Approach to Participatory Development*, Penang: Southbound.

13. Bessette, Guy (2004) *Involving the Community: A Guide to Participatory Development Communication*, Penang/Ottawa: Southbound/International Development Research Centre.
14. Bessette, Guy (2006) *People, Land and Water: Participatory Development Communication for Natural Resource Management*, London/Ottawa: Earthscan/International Development Research Centre.
15. Chin, Saik Yoon (2006) Facilitating participatory group processes: Reflections on the participatory development communication experiments, in Guy Bessette Guy *People, Land and Water: Participatory Development Communication for Natural Resource Management*, London/Ottawa: Earthscan/International Development Research Centre, pp. 273–285.

3
A Synergy of Gross National Happiness and Sufficiency Economy: Implications for Development Communication in Sustainable Social Change

Boonlert Supadhiloke

Introduction

The world community has been recently hit hard by a series of international events and crises, ranging from the global financial crisis, the 9/11 terrorist attacks on the United States, climate change and pandemics. Such unfortunate incidents have undoubtedly dealt a severe blow to the development efforts all over the world. It is imperative that nations across regions join hands to come up with right strategies and tactics to tackle these formidable problems.

The *Happy Planet Index* (HPII) may be cited as a good example of achievement in the new direction of national development through international cooperation. HPII marked a shift in the development paradigm from economic growth to human well-being or the human development index in the wake of the global financial crisis.

In line with HPII, some developing countries have embarked on their own initiatives to gear up development efforts toward set goals. The Buddhist Kingdoms of Bhutan and Thailand have set milestones in the development history with the implementation of *Gross National Happiness* (GNH) and *Sufficiency Economy* (SE). The two Asian countries have some unique characteristics in common that have contributed to the new development approach.

The purpose of this study is two-fold:

1. To analyze and synthesize GNH and SE philosophies as an alternative development paradigm toward the goal of human well-being.
2. To draw up implications for development communication in sustainable social change.

Methodologically, the study is based mainly on analysis of secondary data from various sources, particularly research reports, official and legal documents, articles, periodicals and newspapers. In addition, case studies have been conducted based on personal observations and field work to illustrate the potential of communication in the sustainable development process.

Overview of gross national happiness

Except for the geographical areas and sizes, Bhutan and Thailand share some unique characteristics. Both Asian countries have a monarchy and Buddhism is their national religion. The two institutions are found to exert a strong influence on the development of GNH and SE philosophies.

Bhutan is a small, landlocked country high in the Himalayan mountains. Over half a million people live in this tiny kingdom, squeezed between the world's two most populous countries, India and China. The tiny Himalayan kingdom has been under the Wangchuck dynasty since 1907 with Jigyel Ugyen Wangchuck being the first king. Jigme Dorji Wangchuck, the third king, was highly credited for leading the nation to modernization and upgrading to a high standard the living conditions of the people. He is recognized as the father of modern Bhutan (Servaes, 1994).

However, it was the fourth king of Bhutan, King Jigme Singye Wangchuck who, in 1972, coined the philosophy of GNH. Envisioned as the ultimate development goal, GNH is the guiding philosophy behind every development initiative, encapsulated in every activity and program implemented by the various agencies of the government, and the corporate and private sector. The philosophy strikes a balance between economic, social, spiritual and cultural needs of the people and is considered more important than just increasing and achieving the indicators of gross domestic product (GDP) (Drukair, 2012: 32).

During the 1970s, when several industrialized countries were hard hit by a global economic crisis, Gross National Product (GNP) was considered to be a "broken promise". Prime Minister Jigme Thinley of Bhutan once described "Greed, insatiable human greed" as the cause of today's economic catastrophe in the world (Mydans, 2009). GNH was seen as a good change away from the economic-based GNP.

Under the new constitution, adopted in 2008, government programs – from agriculture and transportation to foreign trade – must be judged not by the economic benefits they might offer but by the happiness they produce for the inhabitants of the tiny mountainous kingdom of Bhutan.

In short, the people-centered development and the theme of "One Nation, One People" laid the foundation for the GNH philosophy during the reign of the fourth king.

King Jigme Singye Wangchuck was so popular and well loved by his people that he was named "King of People". He set milestones in Bhutanese history. On 17 December 2005, a national day, the King announced he was stepping back from power and gave the throne to his son, Jigme Khesar Namgyel Wangchuck, who was crowned in November in the new role of constitutional monarch without executive power. The move came as a shock and against the will of many of his beloved citizens.

The Bhutanese produced an intricate model of well-being that featured the four pillars, the nine domains and the 72 indicators of happiness (Khampa, 2009).

Specifically, the government has determined that the four pillars of a happy society involve the economy, culture, the environment and good governance. It breaks these into nine domains:

1. Psychological or emotional well-being.
2. Standard of living.
3. Health of an individual.
4. Education.
5. Eco-system diversity and vitality (ecology).
6. Cultural diversity and vitality.
7. Time use and balance.
8. Community vitality.
9. Good governance.

Finally, it is noteworthy that on 2 April 2012, Bhutan hosted a pivotal event at the United Nations (UN) in New York in which some 600 people, among them world leading scholars, entrepreneurs, academics and visionaries from the fields of economics, business, politics, environmental and social science, gathered to discuss a challenging topic of Bhutan's commitment to GNH in developing its country and economy. The meeting was intended to boldly take this philosophy to a further level by determining what it will take to develop a new economic paradigm for a world of human happiness and well-being (Drukair, 2012: 24).

Sufficiency Economy philosophy

The philosophy of SE was originally conceived by His Majesty King Bhumibol Adulyadej in 1974. However, it was not until 1997 that the philosophy became widely publicized when the global community encountered a severe financial crisis and Thailand had to devalue its currency and obtain loans from the International Monetary Fund (IMF). His Majesty mentioned a remedial process in his address on the eve of his birthday on 4 December 1997.

> Recently, so many projects have been implemented, so many factories have been built, that it was thought Thailand would become a little tiger and then a big tiger. People were crazy about becoming a tiger.
>
> Being a tiger is not important. The important thing for us is to have a sufficient economy. A sufficient economy means to have enough to support ourselves.
>
> A careful step backward must be taken. A return to less sophisticated methods must be made with less advanced instruments. However, it is a step backward in order to make further progress.
>
> (Government Public Relations Department, 2007: 101–102)

Later, in his royal speech on 4 December 1998, His Majesty clarified the SE philosophy:

> This "sufficiency" means a "sufficiency economy." It is satisfactory if each individual has enough to live on, but it is even better if the entire country has enough to live on.
>
> We in Thailand used to be sufficient. Now, it is free to be insufficient. Therefore, it is a policy to adopt a sufficiency economy so that everyone has enough to live on. Sufficiency means to lead a reasonably comfortable life, without excess or overindulgence in luxury but having enough.
>
> (Government Public Relations Department, 2007: 102)

His Majesty also explained that SE differs from the English term "self-sufficiency":

> Self-sufficiency means producing enough for one's own need, without having to borrow anything from others, or as they say, standing on one's own feet. But the Sufficiency Economy concept is much broader than that. Sufficiency is moderation. If one is moderate in one's desires, one will take less advantage of others.
>
> If all nations hold this concept, without being extreme or insatiable in one's desires, the world will be a happier place.
>
> Being moderate does not mean to be too strictly frugal: luxurious items are permissible, but one should not take advantage of others in the fulfillment of one's desires. Moderation, in other words, living within one's means-should dictate all actions. Act in moderation, speak in moderation that is, be moderate in all activities.
>
> Therefore, sufficiency means moderation and reasonableness.
> (Government Public Relations Department, 2007: 102–103)

The Office of the United Nations Development Program (UNDP) in its 2007 Human Development Report hailed SE as a valid approach to development because it offers a way to avoid "growth that is jobless, ruthless, voiceless, rootless and futureless". The report concludes:

> The Sufficiency Economy offers a way to avoid mindless growth through application of a set of principles which can seem disarmingly simple, but which are rooted in observation of the real world, and underwritten by humanist theory. The appeal of the approach is that the principles are easy to grasp, but the daunting fact is that the application demands high standards of commitment and integrity. But then, if development were easy.
> (Government Public Relations Department, 2007: 103)

To sum up, SE is the philosophy that addresses the way of living and practice of the public in general, from the family unit and the community to the national level in development and management of the country toward the middle path, especially in developing the economy to keep up with the world in an era of globalization. The word "sufficiency" means *moderation, reasonableness* and the need to have *self-immunity* to be ready against any internal and external shocks.

According to the office of the National Economic and Social Advisory Council (NESAC) (2007: 5), SE should mean *self-reliance*, with the aim to utilize local resources as much as possible, while not damaging existing natural resources and the environment. It should regenerate resources and the environment or enhance all forms of capital including human capital, which comprises perseverance, capability, *sati* and *paññа* (mindfulness and wisdom), and mental development; the social capital which is made up of generosity and willingness to assist and honesty leading to the collaboration for the benefit of oneself and society; and the physical capital which includes financial capital, machines, equipment and various technologies. In order to achieve such a situation, it would require the society to be without too much greed or selfishness that would allow everyone to live in moderation, that is, to have enough to live for and to live on, or have sufficient means to survive. Also, citizens of the society must be aware of and take the precautions in all of their decisions. All these efforts would help build self-immunity, especially in developing the nation, where the development must be carried out in step and with the use of appropriate technical knowledge at every level.

As in the case of Bhutan's GNH, questions have been raised about the measurement of SE. It can be seen that the outcome of SE, happiness, can occur at all levels of the holistic system, from an individual, family, community or organization to society, a nation and the world. Attempts have been made to formulate sets of indicators for SE at various levels. Noteworthy is the pioneering work by the office of the NESAC, which constructed a set of indicators based on the philosophy of SE in order to evaluate the macro-performance of the Thai government, from 2005. In the process, the group of researchers breaks down the conceptual framework into eight domains or dimensions as follows:

1. Economic.
2. Ethical and cultural.
3. Social.
4. Education.
5. Management of the public sector.
6. Management of the private sector.
7. Science and technology.
8. Resource and environment.

Each dimension contains a set of items or indicators which are later combined into a composite weighted *Sufficiency Economy Index* or in the so-called *TERMS model of Rural Community Self-Reliance* (Sanyawiwat,

2003). The integrated framework, TERMS, stands for technology, economic, natural resource, mental and socio-cultural. This model views the self-reliance of a community as a goal of community development. It is the result of extensive research, which the Thai National Research Council commissioned to the Science and Technology Institute. More than 50 academics, from governmental bureaus, universities, the private sector and community leaders from five villages were involved. It took them more than seven years to arrive at what now is being called the *Thai concept of community development* (for more details, see Servaes & Malikhao, 2007).

Implication for development communication

The preceding analysis of Bhutan's GNH and Thailand's SE clearly indicates that the national development paradigm has shifted its goal from GDP to GNH or Gross Domestic Happiness (GDH) which is in line with the HPII (Pongpis, 2007). This paradigm shift is significant in that the index of development is not mainly based on an economic foundation but on other factors like psychological well-being, environmental use, human and socio-cultural capitals and good governance as well.

A close examination of both GNH and SE reveals one missing component which is quite essential in a development process. This is the human communication factor which has not been explicitly spelled out in the Bhutanese and Thai paradigms, although it may be implicitly assumed in the processes.

It is the basic assumption that the shift of development goal from material growth to spiritual or mental growth would not be achieved unless the root cause of the problem has been tackled. That root cause of the world's economic catastrophe has been earlier spelled out by the Prime Minister of Bhutan as "greed, insatiable human greed". SE is based on Buddhism's Noble Eightfold Path (*Ariyamagga*) or the Middle Path toward the goal of perpetual happiness. To reach the goal, all human beings must strive to free themselves from such defilement or *kilesa* as craving, anger and illusion – all causes of mental suffering. It is quite interesting that the Noble Eightfold Path has spelled out the concept of communication known as a "right speech" or *Sammavaca* which has been interpreted to be closely related to the concept of "right to communicate" as a basic human right (Supadhiloke, 2007: 325). Thus, from Buddhism's perspective, communication is an essential instrument to reach the goal of happiness in life. It is imperative to integrate communication in the synergy paradigm of SE and GNH.

In Buddhism, the right speech is a broad and multi-dimensional term. At this juncture, it suffices to explicate the concept as a basic human right to indicate an individual's rights to freedom of expression and freedom of information. Of particular importance, Buddhism's Five Precepts spells out one element: not to lie, or to tell the truth. Truthfulness has become an ethical value among Thai people and journalists as a way of life. In addition, the right speech is interpreted to cover not only telling the truth, but also to avoid idle, rude and inciting talk. The right speech can also be seen to encompass cultural values and beliefs. For example, Thailand has been characterized by foreign experts as being collectivism, femininity (Hofstede, 2001: 215) and a high-context communication system (Hall, 1976: 98).

Obviously, Buddhism has a great influence on human communication that, in turn, can further influence the development of human well-being and a happy society. Of particular significance is the implication of the right to communicate, encompassing freedom of expression and information for a democratic and civil society. That free media can promote good governance and democracy is a well-accepted idea. Amartya Sen (1981), the Nobel Prize-winning economist, has argued forcefully since the early 1980s that no famine has ever occurred in any independent country with a democratic government and a free press.

A number of studies show that Buddhism has added a "moral and ethical component" as the fourth dimension of the concept of "communication competence" (e.g., Vasuratna, 2010; Sriussadaporn-Charoenngarm & Jablin, 1999). The other three dimensions are generally known as affective, behavioral and cognitive components (Vasuratna, 2010: 139). The fourth precept of Buddhism, "refrain from telling lies", constitutes a good moral or ethical principle for an individual. Buddhism teaches that one should focus on morality rather than material things. Buddhists are taught to value what a person has done in good deeds, rather than valuing the amount of money a person earns. "To avoid doing evil deeds, to cultivate good deeds and to purify one's mind" is the core value of Buddhism.

Moreover, Buddhism teaches Thai people to be "mindful" at all times (right mindfulness). Doing things in an emotional manner is not mindful. Wisdom (or *paññā*) will come only when one is mindful (or *sati*).

All said, this brief analysis has shown that Buddhism has provided not only a conceptual framework for the philosophies of GNH and SE but also a guideline for human communication conduct. Subject to further empirical studies, such communication components can be formulated as an alternative development strategy to guide the new spiritual growth paradigm.

Case studies

The following case studies are designed to illustrate the potential of communication in the implementation of SE and GNH. Although communication has not been explicitly integrated in the two development systems, it was found to be implicitly inherent in the development implementation and, in many cases, fared satisfactorily to help achieve the development goal.

The first two cases are intended to present the qualitative analysis of communication functions in the SE-related projects, while the third case presents empirical evidence of communication effects in the GNH community.

Case 1: Integrated agricultural farming

The first case study took place at Ban Trok Fai Mai, Klaeng District of Rayong Province, where the local people were led to apply the SE philosophy in an Integrated Agricultural Farming project. Development communication, in this case, focuses on the communicator or opinion leader who has been well oriented by Buddhism's teaching and practice. Elements of participatory communication further supplement a two-way interaction process between opinion leader and followers within the folk cultural context, combined with conventional local media. The case also plays up a network of interpersonal communication and information exchange within and between communities.

The opinion leader of the community, named Mr. Arh Promethaisong, is quite unique and outstanding in that he had been addicted to smoking, drinking, gambling and other social evils before becoming an accomplished communicator. His lifestyle changed when one day he was taken to stay at the Asoke community in Nakorn Prathom province, north of Bangkok. Arh stayed at the Asoke community for four months and four days, undergoing a series of extensive orientation courses and training programs. After completion of the training courses, he returned to work at his own community. "Before our departure from the Asoke community, we were asked to sit down before the picture of His Majesty the King and pledged that we would be loyal to and do good things for our society", Arh said on 18 March 2008.

It is noteworthy here that Asoke communities were among a few communities throughout Thailand that have strong indigenous culture and have adopted a firm stand against capitalism and consumerism. As a result, they did not fall prey to the effect of the 1997 financial crisis.

The Asoke community came into existence after 1975, when its founder, Phra Photirak or Samana Photirak, a former accomplished television producer, entered the monkhood at Wat Asokaram Temple, Samutprakarn Province. As a disciplined and self-restrained person, he commanded great respect from numerous people, both monks and ordinary persons. The Asoke community has been expanded into networks throughout Thailand.

The network of Asoke communities observes the simplest form of living. Being vegetarians, they are very disciplined and adhere to the Five Precepts of Buddhism:

1. abstain from taking life;
2. abstain from stealing;
3. abstain from sexual misconduct;
4. abstain from false speech; and
5. abstain from intoxicants that cloud the mind.

They have followed a self-reliant mode of existence which requires only a small amount of investment, based on local resources and socio-cultural capitals without incurring any external debt. The Asoke community based their way of life on the integration of three components: home, temple (religious organization) and school. They have also operated their own network of mass media and Internet as channels to communicate with their members and publicize their activities. The SE philosophy has been integrated into their self-reliant mode of living and is being implemented in many parts of the country, for example, Ubon Rajathani province.

During his study at the Asoke community, Mr. Arh Promethaisong learned a great deal about SE philosophy and put it into practice after returning to his own community. He began his new life as a small farmer, engaged in integrated farming of rice, vegetable, fishing and raising pigs and poultry. Based on the principles of moderation and reasonableness, he used all existing local resources and folk wisdom in farming. For example, he made use of locally produced organic fertilizers, instead of chemical fertilizers, and raised pigs on waste rice and vegetables.

Having been a successful farmer, Arh has become a community leader as well as a change agent who had to convince other fellows to adopt the practice of SE. This is tantamount to empowerment of the local people. Arh first formed a core group of community leaders who later secured the participation of community fellows through a series of meetings, discussions, exchange of ideas, trainings, organization of demonstrations and social events and other community activities and so on. Usually, Arh and his core group held regular meetings and discussions at a simple public place, called Sala Thai, where one can also observe the picture of His Majesty the King as a source of spiritual inspiration and a symbol of unity. In addition, he often held a series of demonstrations, trainings and observation tours at his own farms and house so that visitors could see by themselves the output and outcome of the sufficiency economy applications. "Our job is really a frightful sweat. With His Majesty the King as our spiritual symbol and role model, we were not discouraged and were determined to succeed. Sufficiency Economy is the right answer in our community", said Arh and his core group at a meeting on 18 March 2008.

Another interesting aspect of participatory communication is the network of interpersonal communication channels at the community level. Since he is a member of the Asoke community in Nakorn Prathom, Arh has become a "bridge" to other Asoke networks throughout the country. At the same time, with his sociable character and communication skills, Arh has also established a close "liaison" with many other community networks, for example, the Youth Group for Khao Chamao Environment Conservation and the nearby Saving Group. This web of local networks is important in Thai culture for the transfer of folk wisdom and knowledge in the participatory development process.

Another unique characteristic of the participatory communication at this community is the existence of community broadcasting known as Broadcast Tower and Community Radio. Although they remain under the control of some community leaders and do not really reflect the voices of the grassroots, both Broadcast Tower and Community Radio play a significant role in the community-building process, as explained by Mr. Arh on 18 March 2008.

Case 2: Youth network for environmental protection

Some successful cases of Thai youths related to the SE philosophy exist as well. Noteworthy is the Youth Group for Khao Chamao Environment Conservation at Tambon Thoong Kwai Kin at Klaeng District of Rayong Province, which lies about 100 km from the capital city of Bangkok. The community leader is a young lady named Buphathip Chaemnil, nicknamed "Pa Fab", a native of Thoong Kwai Kin community. After having completed her secondary education in the community, she went to Bangkok to further her undergraduate studies at Ramkamhaeng University, where she was well known for her extracurricular activities. After graduation, Buphathip returned home to work in the community. She set up her own "rent" book store called "*Namchai*" (heart spirit) where children and youths in the community gathered together and engaged in readings and other social activities.

As a social activist, Buphathip led groups of children to carry out community activities such as growing trees, collecting garbage and cleaning public places. In 1994, she began to lead the children to clean the nearby Khao Chamao National Forest Park where a large number of visitors from various parts of the country came for recreation every year. Since then, these young volunteers from Khao Din Village have regularly frequented the Khao Chamao mountainous areas and erased dirty marks in the caves, grown trees and prevented the forest from being illegally cut and burnt. The number of these young volunteers has gradually increased year by year from about ten persons in 1994 to around 400 persons in 2009. The group became more organized and called itself "Khao Chamao Conservation Group" (Kloom Raksa Khao Chamao – in Thai). It expanded into networks within and outside communities. Their activities have also been expanded to cover, for example, taking care of birds and elephants. The local network has since developed into a regional network of Thai youths with the primary purpose of environmental protection and conservation.

All the young people were born in and lived in the vicinity of the Khao Chamao forest park. They have been deeply embedded in the indigenous culture and learned to treasure these natural and cultural aesthetics and values. For them, trees are their life and the environment is a most precious natural capital for development. Out of these common bond and concern, the young people were moved to join hands to work for the community.

Viewed from the SE perspective, the youths at Ban Khao Din village were found to lead a moderate way of life in harmony with their natural environment. Based on the group decision, the youths were determined to transform their existing natural, social and cultural capitals into sustainable community-based development. By social, cultural and natural capitals, they mean people's strength, aspiration and knowledge, ethical integrity, folk wisdom and environmental resources. They also adhered to Thai Buddhist preaching and cultural values by "abstaining from, reducing and giving up all forms of social evils such as gambling, alcoholic drinking, violence, sexual and drug abuse which are considered a primary cause of poverty".

Apart from Buddhism's Five Precepts, the Noble Eightfold Path has been taken by people as a guiding light for the conduct of their self-reliant livelihood. As mentioned earlier, the SE philosophy has become widely applied in local communities after the year 1997 when the Thai economy was hard hit by the globalization effects of the world financial crisis. The Thoong Kwai Kin community had also been threatened by the industrialization and globalization processes which the people called "storms from the West". Several villagers said that Thailand had adopted the Western culture for so long that Buddhist culture had been forgotten. The Western concept is based on capitalism, which emphasizes acquiring wealth and material products to satisfy human needs and greed. It is an illusion of the world, based on materialism and consumerism. Man is driven to possess material wealth as much as possible to satisfy his endless needs and become obsessed with self-interest and lust. Thai Buddhist culture, on the other hand, emphasizes spiritual growth and happiness, based on such socio-cultural capitals as folk wisdom, knowledge, ethical integrity and compassion.

In every Thai community, a senior monk is an important opinion leader who commands respect from local people and imparts Buddhist preaching for their spiritual development. The Middle Path or moderation, a core process of the SE, also originates from Buddhism's doctrine, named the Noble Eightfold Path, which seeks a balance between ethical and knowledge inputs and happiness outputs for human life. The senior monk at Thoong Kwai Kin Temple, Phra Kru Prachote Dhammapritome, explained that individual happiness is not based on one's physical needs like eating, sleeping or sex; it is rather based on moral virtues such as offering merits and helping needy people. One common village practice, named "Karn Long Kaek" or "Mutual and Communal Help", has been a long-standing folk wisdom or custom in Thailand by which people come to help one another in farming or other community activities, using their own social capital, regardless of financial costs.

The youths set out to formulate the "Happy Community Project" based on community data, aimed at achieving happiness in life. The core strategy was "less dependence on outside resources" but "more self-reliance on existing socio-cultural and human capitals". They ventured into the "integrated community development approach", based on home, temple (religious place) and school – the three main social institutions in the community. As for communication, the youths selected their leaders as a group of "prime movers" to secure cooperation and participation from people in the community. First of all, these prime movers would seek advice and information from a number of local or opinion leaders; most important among them are Buddhist senior monks, village headmen, district officials, teachers, local pundits or learned men, artists and senior citizens. In Thai culture, with community leaders' approval, it becomes easier for the youths to recruit other people's participation. In the case of the Happy Community Project, a number of social events, activities and meetings were organized at the community learning center at Ban Khao Din in which various groups of local people were invited to take part. Some folk culture and media, such as puppet show dramas and songs, were frequently used to attract local people. Owing to their common historical and cultural heritages, both leaders and community fellows found it congenial to reach agreement on the application of the SE project.

Case 3: Development communication for Bhutan's gross national happiness

The third case presents empirical research on the role of participatory development communication in GNH at a Bhutanese community. It was conducted by a graduate student from Bhutan, Khampa, for his Master's thesis at Bangkok University in 2009 (See, Khampa, 2009). Khampa choose a rural community in Eastern Bhutan, named Khaling Gewog of Trashigang District. Most people there are illiterate farmers. The randomly selected sample of 334 respondents consisted of mainly adult females on a low income, who did not have access to mass media as a source of information. Most villagers relied on face-to-face communication, interpersonal channels and activities, and traditional media in their daily lives. Accordingly, Khampa conceptualized the concept of development communication in "participatory" and "action" communication dimensions, each of which was measured by a set of items reflecting interpersonal, group and cultural activities.

To measure the people's happiness (GNH), Khampa used a short version of the 52-page questionnaire developed by the Center for Bhutan Studies. The shortlist of the questionnaire consisted of 21 items designed to tap the nine dimensions of GNH as earlier described, namely, emotional well-being, standard of living, health of an individual, education, eco-system diversity and vitality, time use and balance, community vitality and good governance. However, after having computed the factor analysis to form the GNH index, only 17 items were found to have significant loadings on eight dimensions. The items designed to measure the dimension of eco-system diversity and vitality (ecology) were found to correlate with cultural and educational items and, thus, this dimension was subsumed under other dimensions.

What is interesting is that in this study the two communication variables, participatory communication and action communication, were found to have a significant relationship with the GNH index. Even after having controlled for the socio-economic status variable, these relationships remained statistically significant. The participatory development communication was found to correlate with GNH at 0.289 ($p < 0.01$), while the communication action correlated with GNH at 0.234 ($p < 0.01$). After having controlled the socio-economic status variable, the two correlations dropped to 0.283 and 0.227 ($p < 0.01$), respectively.

This empirical evidence suggests that communication activities can exert an influence on Bhutanese people's ways of life in remote areas. Given the fact that farmers are poor and illiterate, they can rely on various means of communication, including traditional, interpersonal and group channels and indigenous culture, to help improve their standard of living and well-being. Thus, in the Himalayan kingdom of Bhutan, communication as a social process proved to be an imperative component that empowered the grassroots to lead a happy life in harmony with their socio-economic and cultural environments.

Conclusions

This study has analyzed and synthesized Bhutan's GNH concept and Thailand's SE philosophy as an alternative paradigm for sustainable development and social change. Both development paradigms have been conceived by Their Majesties the Kings of the two countries based on the Buddhist doctrine and geared to achieve human happiness, rather than economic growth. Conceptually and operationally, both paradigms have several features in common. GNH possesses four pillars of a happy society broken into nine domains, namely: psychological well-being, ecology, health, education, culture, living standard, time use, community vitality and good governance, each with its own index. Similarly, SE breaks down into eight domains, namely: economic, ethical and cultural, social, educational, management of public and private sectors, science and technology, and resource and environment, each with its own indicators and index. Accordingly, a synergy paradigm of human communication and happiness has been drawn up to serve as a conceptual framework for study of sustainable development and social change toward the goal of human happiness or spiritual growth.

In the synergy paradigm, communication is suggested as an integral part of the SE, also derived from Buddhism's Noble Eightfold Path, namely, the right speech or *Sammavaca*. The new dimension of communication may be interpreted as the right to communicate encompassing freedoms of expression, information, participation, privacy and such ethical values as truthfulness, accuracy, fairness, and decency; they can also be taken as human and social capitals in the human happiness system. Based on the three case studies in Thailand and Bhutan, the communication components were found to correlate significantly with SE and GNH indices.

In view of the limitations of existing empirical studies, further research is needed to verify the findings and the proposed synergy paradigm.

Notes

During the course of compiling the case studies, the following persons have been interviewed:

Arh Promethaisong, at Ban Trok Faimai, 18 March 2008.
Buphathip Chaemnil, at Ban Khao Din, 20 March 2008.
Cholakant Polwatana, at Ban Khao Din, 18 March 2008.
Orathai Sookthavorn, at Ban Khao Din, 18 March 2008.

References

Drukair (2012) "The Business of Gross National Happiness" *Tashi Delek*, Sep–Oct 2012, 24–32.

Government Public Relations Department (2007) *King Bhumibol Adulyadej: The Great King of Communication* (Bangkok. Paper House Limited Partnership). pp. 101–103.

Hall E. T. (1976) *Beyond Culture* (New York: Anchor): 98.

Hofstede G. (2001) *Culture's Consequences: Comparing Values, Behaviors, Institutions, and Organizations Across Nations*, 2nd edn (Thousand Oaks, CA: Sage), p. 215.

Khampa (2009) *Participatory Development Communication and People's Happiness in a Bhutanese Community*, Unpublished Master's thesis, Bangkok University, Bangkok, Thailand.

Mydans S. (2009) "Recalculating Happiness in a Himalayan Kingdom" *The New York Times*, 6 May, http://www.nytimes.com

National Economic and Social Advisory Council (NESAC) (2007) *Formulation and Evaluation of Sufficiency Economy Indicators and the Economic and Social Report on the National Performance in the Direction of Sufficiency Economy* (Bangkok, Ubon Rajathanee University).

Pongpis S. (2007) *Sethakit Porpiang Kerd Dai Tha Chai Prathana (in Thai) (Sufficiency Economy: To Take Place if Desired*) (Bangkok: Charoenvitya Printing).

Sanyawiwat S. (2003) *Theories and Social Development Strategies* (Bangkok: Chualalongkorn Publishing House), pp. 99–104.

Sen A. (1981) *Poverty and Famines: An Essay on Entitlement and Deprivation* (Oxford: Clarendon Press).

Servaes J. (1994), "Broadcasting in Bhutan: Between tradition and modernization", *Asian Journal of Communication*, 4(1), 111–133.

Servaes J. & Malikhao P. (2007), "Communication and Sustainable Development", Servaes J. & Liu, S. (eds.), *Moving Targets. Mapping the Paths between Communication, Technology and Social Change in Communities* (Southbound: Penang), pp. 11–42.

Sriussadaporn-Charoenngam N. & F. M. Jablin (1999) "An Exploratory Study of Communication Competence in Thai Organizations". *Journal of Business Communication*, 36(4), 382–418.

Supadhiloke B. (2007) "Right to Communicate, Media Reform and Civil Society in Thailand", *Communication for Development and Social Change*, 1(4), 323–328.

Vasuratna V. (2010) *Defining the Components of Communication Competence from the Perspective of Thai Employment Interviewers*. Unpublished doctoral dissertation, Bangkok University, Bangkok, Thailand.

4

Environmental Communication from the Fringes to Mainstream: Creating a Paradigm Shift in Sustainable Development

Kiran Prasad

"The earth provides enough to satisfy every man's need but not every man's greed."

Mahatma Gandhi

Introduction

Most of the agrarian countries of the developing world, including India, have concentrated on economic issues, leading to the neglect of the social, political and cultural contexts of development (Bottomore, 1975). Many countries in Asia, including India, have adopted development patterns based on Western conceptions of modernization and development that have eroded indigenous ways of life and ecological traditions. In the course of development, what attitudes different societies assume toward the environment are determined by the value systems and political processes of decision-making that operate in their respective cultures. All societies are founded on religious, cultural and social mores which govern their attitudes, values and lifestyles. But these fundamental structures have been pushed to the background in the competition for rapid economic growth, mass production and a rich lifestyle. In the race among nations for fast economic growth and prosperity, there has begun a serious introspection where development ends and destruction begins in the context of the natural environment.

Environmental activism entered the political domain in the 1980s. The United Nations (UN) had declared 1993 the International Year of the Indigenous People. This drew the attention of the highly developed industrial societies to the worldview of the indigenous people about Mother Earth being an inalienable inheritance of humankind and their practice of environmental conservation for sustainable living. Women have played a key role in environmental conservation in the developing world. But their activism was viewed as local action and confined to the fringes of mainstream development. The Millennium Development Goals (MDGs) that regard empowering women and environmental sustainability as the key factors for development began to recognize the international role of women in protecting and sustaining the ecology. These goals demand a renewed look at indigenous models of living in which women play an important role in environmental conservation. The existence of eco-religious practices among the people of India points to the long-standing traditions of ecological conservation and a culture of nurturing nature. These traditions on the fringes are undergoing a transformation to emerge as a mainstream political ecology that is gaining currency in the development discourse.

Women have played a significant role in effecting a paradigm shift in political ecology in which environmental conservation will be the touchstone of sustainable development. Often excluded from the mainstream media, women are using alternate media in green communication.

This chapter will focus on women's interventions in environmental communication through the mainstream and alternative media and also women's activism in mainstream environment communication in development in the context of the emerging political ecology of India. This chapter also highlights the case of the Narmada Bachao Andolan movement against big dams, spearheaded by the veteran environmentalist Medha Patkar, as it completes 25 years of its struggle to assert the rights of the people to their land, water and forests.

Environment and sustainable development

Harold Lasswell's definition of politics as to who gets what, when and how, *and at what expense* (added by Mr. Harish Khare, Advisor to the Prime Minister of India) can be very much applied to the realm of development as well. This sums up a complex interlinking of the environment with development in different countries. Economic growth and development has adversely impacted the environment in some regions, whereas environmental degradation has adversely affected economic growth and development in some regions. The former trend is clear in the developing countries of Asia, while the later trend is visible in Africa. The United Nations Human Development Report (2010) has calculated that South Asia is home to half of the world's multi-dimensionally poor population of 844 million people. Eight states of India are home to 421 million multi-dimensionally poor people, which is more than 410 million in the 26 poorest African countries.

The resource degradation of the past increases today's poverty, while today's poverty makes it difficult to care for the future. It is evident that environmental problems, depending on the degree of development and the structure of the economy, development and the environment, are global issues. Some resources may be shared globally, but there are varying levels of resource consumption patterns and economic policies in different countries. This has led to a situation where environmental problems existing in some countries may be linked to the practices and policies in others. For instance, in a global economy the benefits of resources, knowledge and environment policies in developing countries often accrue to developed countries and at the same time technologies needed by developing countries often lie with the developed countries.

Development paradigms have relied heavily on top-down approaches with information, knowledge and expertise flowing from scientists, development planners, policy makers and bureaucrats, who advocated the lab to land approach symbolized by the green revolution in India. As Vandana Shiva puts it:

> The so-called green revolution was neither green, nor revolutionary. It has dispossessed small peasants, pushed our rich agro biodiversity to extinction, mined our aquifers, desertified our soils and undermined our nutrition and health ... It was supposed to create prosperity, but it left farmers steeped in debt. Both in financial and ecological terms, industrial agriculture and chemical farming is based on a negative economy – it uses more inputs than it produces. The consequence is impoverished eco-systems and impoverished and indebted farmers.
>
> (Shiva, 2009: 25)

Farmers, tribes and women, previously viewed as at the fringes of this top-down approach, have emerged as innovative farmers who challenge many mainstream notions of commercial agriculture. Agricultural innovation has entered a new phase, with organic farming and a reverse strategy from the land to the lab where agricultural scientists go to the field and often return to the lab to understand the regenerative power of nature and the knowledge inherent in communities.

Eco-religion in India

Global concerns over the state of the environment saw the UN declaring 1993 as the International Year of the Indigenous People. This drew the attention of the highly developed industrial societies to the world-view of the indigenous people about Mother Earth being an inalienable inheritance of humankind and their practice of environmental conservation for sustainable living. It is now accepted that contemporary legislations, which strictly curtail the access of forest dwellers to their habitat and reserve forest lands in the name of modern techniques of scientific conservation, have actually thrown out the original conservators of the forests, thus paving the way for their commercial exploitation. The International Chapter of the Indigenous People in 1992 declared that all policies toward forests must be based on a respect for cultural diversity and for promotion of indigenous models of living.

Ancient Indians developed many ideas, attitudes and practices which favored the maintenance of ecological balance for the welfare of all living beings. These views are reflected in several philosophical and religious texts of the *Vedas* and *Puranas,* which form the basis for an environmental ethic. The universe, in *Advaita* philosophy, acquires a cosmic character as it considers all living beings to be the creation of God. This doctrine provides the philosophic basis for the Indian veneration of the natural world. This leads us to conclude that the Indian tradition has an ecological conscience (Crawford, 1982: 149–150).

Indians laid down traditions, customs and rituals, to ensure that the complex, abstract principles they had developed could be put into practice. These practices, over time, developed the technology of agriculture, methods of environmental protection and knowledge of medicinal properties of trees (Banwari, 1992). It is estimated that about 10 per cent of the indigenous tribal population (*adivasis*) in India continue to practice shifting cultivation. A total area of about 50 lakh hectares over 15 states are covered by shifting cultivation in India. The land is not ploughed in this type of farming and neither is there any need to domesticate animals. The cultivators have total confidence in the generative power of the earth and see no need to resort to eco-destructive methods. At the end of summer, the hill-sides are prepared for cultivation by trimming the undergrowth of bushes and shrubs. These are then burnt and the ashes provide the manure. Before the monsoons set in, the shrubs and bushes are set on fire again. As soon as the rains come, the seeds are sown and the earth is activated to produce a rich harvest and this method of farming continues to be practiced in several States of India (Vadakumchery, 1993). Such fringe agricultural traditions of indigenous communities are gaining credence through organic farming and organic products against mainstream commercial agriculture that had pushed ecological concerns to the background in the competition for rapid economic growth and mass production.

Eco-religion of the Bishnois: Biodiversity as a way of life

One of the outstanding indigenous movements on the importance of eco-religion in environmental conservation is that of the Bishnois of Rajasthan, a north-western state in India which has vast tracts of deserts. This is an unusual community, for whom the protection of trees and animals is a religious obligation (Sharma, 1999). The faith that all living things (including animals) have a right to survive and share all resources underlines the basic philosophy of the Bishnoi eco-religion.

They follow a set of 29 rules, which include eight tenets that have been prescribed to preserve biodiversity and encourage good animal husbandry. These include a ban on killing of all animals and cutting of green trees and providing protection to all life forms. The community has even been directed to make sure that firewood is free of small insects before it is used as fuel. Wearing blue clothes is prohibited because the dye for colouring them is obtained from particular shrubs which have to be cut for extracting the dye.

The Bishnois are presently spread over the western region of Rajasthan and parts of Haryana and Punjab. They are more prosperous than other communities living in the Thar deserts, probably because of their eco-friendly way of life. Their villages are easily distinguishable because there are plenty of trees and other vegetation and herds of antelope roam freely near their homes. The fields are ploughed with simple ploughs using bullocks or camels. This causes minimal damage to the fragile desert eco-system. Only one crop of *bajra* (millet) is grown during the monsoon season. The bushes which grow in the fields protect the loose sand from wind erosion and provide the much-needed fodder for animals during a famine. The Bishnois keep only cows and buffaloes, as the rearing of sheep and goats which devour desert vegetation, is taboo. Though they are Hindus, they do not burn their dead but bury them, to save precious wood and trees. They store water all year round in underground tanks by collecting rain water as it is precious in this dry desert area.

Even today, the Bishnois aggressively protect the khejri trees and the antelopes, particularly the blackbuck and chinkara. They consider protecting a tree from the axe, even if it be at the cost of one's head, a good deed. They not only protect antelopes but also share their food and water with them. They keep strict vigil against poachers (a popular actor who was accused of hunting a deer in a Bishnoi village had to face the ire of the local population and was even prosecuted according to the law of the state). It is this environmental awareness and commitment to environmental conservation and protection that makes the Bishnois stand apart from other sects and communities in India (Sharma, 1999). The eco-religious cultural practices of the Bishnois have inspired many women's groups and local communities to take on powerful lobbies that support development based on unbridled exploitation of natural resources and neglect of the environment.

Political ecology and women's leadership

There was a parallel campaign for the right to information during the mid-1990s, which converged with other mass movements such as the movement for women's rights and violence against women, rehabilitation of people displaced by development projects, the movement against big dams and protection of the environment. Non-governmental organizations (NGOs) have played an important role in catalyzing these social movements. The NGOs which were dependent on state support for implementing development programs and were fringe actors have now emerged as non-state transnational and transcultural mainstream actors critiquing development policies and programs. NGOs and people's collectives are paving the way to alternative paths of development. The right to information law in India was an outcome of the convergence of several people's movements.

The groundbreaking experiment in social audit had its genesis nearly ten years ago in the backward districts of Rajasthan, where a mass organization of workers and peasants – the Mazdoor Kisan Shakti Sangathan (MKSS), founded by Aruna Roy in 1990 – began to focus on the rights of the common people to information and made an attempt to break the vicious circle of corruption. The National Campaign for People's Right to Information (NCPRI) was founded in 1996 with the primary objective to campaign for a national law facilitating the exercise of the fundamental right to information. Thus, what began as local campaign almost 15 years ago, later snowballed into a mass movement, culminating in a national law on the right to information in 2005.

The right to information campaign led to a demand for the release to the people of details of all development projects through various channels including the mass media, to initiate open debate and public opinion. The movement aimed to create special-purpose associations and coalitions, public meetings, solemn processions, vigils, rallies, demonstrations, petition drives, statements to and in public media, and pamphleteering. This campaign is a striking one and worthy of emulation in many of the poorer parts of the world, especially since it is the ordinary, illiterate and oppressed rural folk who spearheaded the right to information, as it is integrally linked to the right to livelihood of the poor people.

The mainstream media has been critiqued as commercial in its approach and for not effectively serving local community issues and the interests of the poor. This has led to a search for alternative communication media that would provide greater access to media resources and enable the voices of marginalized communities, particularly rural women, to be heard on their own terms (Prasad, 2008). In this context, women in India have begun to successfully use alternative media to promote gender justice and sustainable development. Community media in various forms – radio, video, television, information and communication technologies such as the Internet and web-based networks – are enabling women to debate about their rights and empower them through knowledge about social, political, economic and environmental policies (Prasad, 2006).

The community radio initiatives in India *Women Speak to Women Project* of the Deccan Development Society (DDS) of Andhra Pradesh; the *Radio Ujjas* of Kutch Mahila Vikas Sangathan in earthquake ravaged Bhuj; *Namma Dhwani*, the VOICES initiative in Budikote, Karnataka; SEWA's *Rudi no Radio* in Gujarat and Agragamee's *Ujjala* in Jeypore, Orissa, have programs designed and produced by women's collectives. *Khabar Lahariya*, a fortnightly rural newspaper in Chitrakoot, Uttar Pradesh, and the magazine *Baini* of the Mahila Samakhya, Nainital, are spearheaded by rural women who bring out newspapers focusing on remote villages that remain untouched by the mainstream media.

The DDS works with about 100 Dalit women's groups, consisting of nearly 4,000 members in 75 villages, on areas such as sustainable agriculture and the empowerment of women. These women were trained in basic film-making by the NGO in collaboration with the United Nations Educational, Scientific and Cultural Organization (UNESCO) and they use their skills to document traditional farming know-how. The DDS has established a media center and its *Women Speak to Women* project has prepared 200 hours of programs which is carried out through community radio and a community video project.

Namma Dhwani (Our Voices), the community radio initiative in Budikote, Karnataka, cablecasts programs made by the community and is aired by the local All India radio station. One of the pioneering community radio programs in India, the half-hour session includes a mix of entertainment and discussion on issues of health, farming, schooling and raising the status of women. Women are actively involved in the management of community radio. The management committee at *Namma Dhwani* comprises ten women, two local men, one representative from the NGO VOICES, and one from MYRADA, a local NGO. The ten women on the management committee are all representatives of the ten self-help groups (SHGs) or micro-credit groups in Budikote, which, between them, represent around 230 women. This committee is responsible for a number of tasks such as the supervision of management and administrative matters; determining program content and providing regular documented feedback and, importantly, recruiting volunteers and encouraging programming by women. The programming schedule also has regular slots devoted to women. The committee has declared every Saturday a "women's zone", where only the women operate the equipment, compare and discuss issues that are relevant to their lives. With programs designed and produced by the women who were relegated to the fringes of media use and production, and in their own idiom, these initiatives endeavour to bring in development from the grassroots while building solidarities at the same time (Prasad, 2006).

Even more surprising is the entry of rural women in the management and production of newspapers given their limited levels of education. Chitrakoot, Uttar Pradesh, is remote and few newspapers reach there. Those that do are in Hindi and don't address local concerns. Women form the editorial team of *Khabar Lahariya*, a fortnightly rural newspaper in Bundeli, the local dialect. It is produced by a group of seven Dalit and Kol women. The group is responsible not only for news-gathering and writing but also editing, photography, illustration, production and distribution. It is supported by Nirantar, a Delhi-based group that works with gender and education. The newspaper was first published in May 2002 and now has a print run of 1,500. It is sold in nearly 200 villages in the district. *Khabar Lahariya* provides a mix of news that touches lives and brings about change. An investigation by the paper into illegal mining rocked the local administration. A wall in the room that acts as the newspaper's office has a framed copy of the Chameli Devi Jain award citation, which the group won for outstanding contribution to journalism (Chakraborty, 2004).

In this transformation of women, from a state of being voiceless and confined to the fringes of social life, there is also a new dimension of environmental activism where women are now assuming leadership in environmental politics. The Narmada Bachao Andolan (Struggle to Save Narmada River) catapulted environmental activism into the political domain in India and has now become symbolic of the global movement against big dams.

The case of the *Narmada Bachao Andolan (NBA)*

Medha Patkar, founder of the NBA and the National Alliance of People's Movements, is a powerful voice for millions of voiceless poor and oppressed people whose uncompromising insistence on the right to life and livelihood has compelled the post-Independence generation in India as well as around the world to revisit the basic questions of natural resources, human rights, environment and development. The NBA began as a fight for information by the tribal and peasant communities about the Narmada Valley Development Projects in Maharashtra, Madhya Pradesh and Gujarat and continued as a fight from 1990 to 1991 for the just rehabilitation of millions of people ousted by the Sardar Sarovar Dam, the world's largest river projects, and other large dams along the Narmada River. The Sardar Sarovar Dam was a project that would displace upon completion 320,000 tribes and submerge over 37,000 hectares of land. When it became clear that the magnitude of the project precluded accurate assessment of damages and losses, and that rehabilitation was impossible, the movement challenged the very basis of the project and questioned its claim to "development". To date, as many as 35,000 people have been relocated by the project; however, they have not been adequately resettled and as many as 40,000 families have returned to their home villages despite the constant threat of submergence and still await resettlement (*The Hindu*, 25 October 2010).

In 1985, the NBA began mobilizing massive marches and rallies against the project; meetings, folk songs and dances formed part of the struggle that asserted the right of the people to their land, water and forests and set a precedent for the subsequent social movements. The NBA also began to go global with the help of the Internet and new media. The NBA website carries press releases, images, online appeals, petitions, court orders, case studies, comments and critiques, and responses of people who advocate alternative approaches to the big dam. As an outgrowth of her work to stop dam construction, Patkar has helped establish a network of activists across the country – the National Alliance of People's Movements. Linking the NBA with hundreds of peasant, tribal, dalit, women and labour movements throughout India, Medha Patkar is Convener of the National Alliance of People's Movements – a non-electoral, secular political alliance opposed to globalization – liberalization based economic policy and for alternative development paradigm and plans. Medha Patkar says:

> If the vast majority of our population is to be fed and clothed, then a balanced vision with our own priorities in place of the Western models is a must. There is no other way but to redefine "modernity" and the goals of development, to widen it to a sustainable, just society based on harmonious, non-exploitative relationships between human beings, and between people and nature.
>
> (*The Hindu*, 2004)

Many of the uprooted families of large development projects such as big dams are tribal people and forest dwellers who are among the most underprivileged and vulnerable people in India. The viability of big dams which submerge adjoining villages was intensely questioned in 2006 with the Tehri dam completely submerging the historic Tehri town in Uttarakhand. Besides, there are rational, cost-effective and environmentally more benign alternatives to the Sardar Sarovar Project (Bidwai 1998). Check dams, rain water collection tanks, and mini water projects, with the active involvement of the local people in several states, have evinced greater support and participation rather than large dams involving massive funds with heavy costs of submergence, environmental impact, rehabilitation and resettlement of the affected people. The Narmada Bachao Andolan (NBA) has brought to light many of these issues associated with big dam projects. It has helped to open public debate and consultation regarding the cost and benefits of such projects, and environmental damages.

The NBA has been able to drive home the message that "There Are Many Alternatives" as against the global media refrain that "There is no Alternative" (TINA). "The NBA is oxygen for other movements, giving them the strength to fight", said the general secretary of the National Hawkers Union, Shaktiman Ghosh (Gaikwad, 2010). Activists, environmentalists, farmers, adivasis[1] and supporters, young and old, thronged the village square in Dhadgaon of Bhadal in Madhya Pradesh for a public meeting, singing "In the Narmada valley/The fight is still on" where Narmada Bachao Andolan (NBA) leader Medha Patkar led a prayer meeting as part of the activities marking the completion of 25 years of the people's movement (Rahi, 2010).

The NBA has inspired many local communities to protest against mining and logging rights given to big commercial business houses. The agitation led by C. K. Janu for asserting the right of the tribal people over forest land in Muthanga, Kerala, has been another long struggle. The recent *Avatar*-style protests against mining rights in Orissa granted to Vedanta by actors in the United Kingdom point to an evolution of transcultural flow of ideas and images that are becoming integral to environmental activism. In Kerala, several protest movements have been launched against the unregulated drawing of ground water by the Coca-Cola and Pepsi companies from Plachimada and Kanjikkode in Palakkad, championed by the *adivasis* led by Mayilamma. The Plachimada struggle attracted global attention when reports of the BBC and the New Delhi-based Centre for Science and Environment confirmed that the Coca-Cola sold by the company contained more than the admissible level of harmful chemicals. Medha Patkar and Vandana Shiva along with several environmental activists and global NGOs arrived in Plachimada and extended their support to the agitation. After long legal battles, the Court ruled that the company had the same right as the people to draw water and share natural resources.

In India's real-life *Avatar,* there is a struggle to save the mountains that are sacred to the tribes and is their only habitat, from a corporate giant, Vedanta from the United Kingdom. The Dongria and Kutia Kondh, the tribes that live in the 250 sq km Niyamgiri Hill Ranges of western Orissa's Kalahandi district, have been officially recognized as primitive tribal groups that need special protection. On YouTube, a series of campaigns has been running against Vedanta. The Niyamgiri Hills, which the Dongria Kondhs worship, are the center of rich deposits of bauxite that Vedanta wants to mine. The Saxena Committee that enquired into Vedanta's business suggested that the Vedanta operations endangered nearly 750 sq km of forest land. The Environment Ministry has also stepped in to keep Vedanta from developing its plans to mine in the Niyamgiri Hills after the tribes won a court verdict in their favor.

It is estimated that 15 million people worldwide earn a living through waste picking. Waste pickers are now seen as offering a solution for climate change mitigation and waste management that makes economic, ecological and social sense. The UN Framework Convention on Climate Change (UNFCCC) Conference in Tianjin, China saw three women rag pickers from India sharing their experiences on the steps taken by them to reduce greenhouse gases. Baidabai Gaikwad from Pune, Maya

Khodave from Nasik and Sushila Sable from Mumbai, were representing waste pickers from around the world at the UN and explained the implications of the UNFCCC process on the marginalized in different parts of the world. These environmental movements led by women, tribes, peasants and local communities are mainstreaming their perspectives to create awareness about the rights of people to local resources, prevent their over-exploitation and promote conservation efforts. Their voices, which were rarely heard or even considered in development policies, have now begun to acquire a new meaning with the local communities assuming custodianship of their environment and natural resources which previously was the sole responsibility of the government.

The Nagoya Protocol, a new international treaty, is seen as one of the victories for India and other developing nations, to ensure that the benefits of natural resources and their commercial derivatives are shared with local communities. The Nagoya Protocol includes a sweeping plan to protect bio-diversity by setting targets for 2020. Nations agreed to make 17 per cent of the globe's land area and 10 per cent of coastal and marine areas into protected regions, as opposed to the current levels of 13 and 1 per cent. They also agreed to bring "natural capital" into national accounting systems, so that the trillions of dollars' worth of benefits that nature provides to economies are valued. This agreement is a result of almost two decades of UN negotiations, where India leads a group of 17 vastly diverse countries with rich reserves of exploitable natural resources.

The new Access and Benefit Sharing (ABS) rules mean that multinational companies will have to share their profits with local communities not only for using the original resource, but also any derivative products developed from it. For example, a pharmaceutical company which develops a new drug from ingredients found in an Indian plant will now have to give a fair share of profits to the Indian communities which nurtured the plant in the first place. International drug firms will also have to pay to use human genetic material such as pathogens – the germs responsible for virus pandemics which are used to develop lucrative vaccines. This is the light of the previous practice of companies who just took the pathogens from the developing countries, made the vaccine, and then sold it back to them for billions of dollars. However, the flip side is that the United States – one of the largest users of such resources – is not among the nearly 200 signatories of the rules of the Nagoya Protocol (Jebaraj, 2010). It reminds us that the development paradigm shift of including fringe voices and perspectives into the mainstream discourse is indeed going to be a long struggle for millions of the poor and disadvantaged against elite development models that curtail their access to natural resources and livelihoods.

The silver lining

There is a growing emergence of webs and networks of relations among individuals, groups, communities, states, international organizations and transnational actors in environmental politics and activism. This is leading to a challenge to global corporations that attempt to have a homogenizing impact on lifestyles and consumption patterns. The right to information campaign led by Aruna Roy, struggle for natural resources led by Medha Patkar, the sustainable agriculture movement led by Vandana Shiva and a host of local agitations led by women like C. K. Janu and Mayilamma have converged and have played a stellar role in grounding political ecology in India. Those who were long neglected and silenced in the development processes, particularly women, have been awakened by people's movements such as the NBA. Owing to networking with the NBA, women in the local collectives/federations have increased contact with the bureaucracy in the government offices. Prior to the networking with the National Alliance for People's Movements, many women were unaware of where the government offices were located and played a minimal role in the political life of the country. Now women representatives elected to the local government bodies visit the state offices and interact with officials and are also informed by them about various development programs. In addition to greater interaction with government officials, networking with national organizations such as the National Alliance of People's Movements (NAPM) helps women take up leadership roles in their own neighborhoods, communities and villages.

Nevertheless, the degree of power that autonomous women's organizations can exert within the political process often determines the outcome in the form of appropriate policy changes for development. However, women's empowerment is generally perceived as a matter of "low politics" and policy reforms and strategies are therefore influenced primarily by the extent to which different policies are likely to influence political stability (Heise, 1994). Now there is a growing acceptance within the women's movement that any strategy to achieve significant policy reforms at the state, regional or local level must engage the state and demand reform. Women who attempt to reform the process of development by resistance against bureaucratic structures often find that an alliance of politicians and vested interests have an almost impregnable hold on the central institutions of the society. The increasing poverty of these women often do not allow them to think leisurely and realize that the rise of popular protest which intends to bring in change must build up from the fringes of society to the center until this becomes the dominant socio-political organization in the country (Chakravarty, 2003:638).

Further, women's extreme poverty and their pseudo-enfranchisement in the developing countries sometimes make them realize that their powerlessness will not easily change their circumstances. This situation is complicated by policy makers who often ignore a world in which women are seldom leaders, doers and supporters of families, and where women's concerns like poverty and lack of health care are absent or trivialized. But the struggle of women, as narrated during the course of this chapter is strongly founded on the hope that status and legitimacy will be bestowed on women's issues. It does give an indication as to the realization of women of the changes in themselves as they prepare to raise difficult questions regarding the environmental sustainability of development projects. If satisfactory answers are not forthcoming, women are gradually moving toward a position of seeking alternative models of development that put people's welfare on the top of the agenda though they still have a long way to go. As Swami Vivekananda said: "Bring light to the poor and bring light to the rich, for they require it more than the poor. Bring light to the ignorant and more light to the educated, for the vanities of the educated are tremendous" (Swami Vivekananda, Vol. III, 1985: 228). I hope we will become active custodians of our environment.

Note

1. *Adivasis* is the collective name used for the many indigenous peoples of India.

References

Banwari (1992). *Pancavati: Indian Approach to Environment* (Tr. Asha Vohra). New Delhi: Sri Vinayaka Publications.

Bidwai, P. (1998). "A Narrow-minded Approach", *Frontline,* 9 October, pp. 105–106.

Bottomore, T.B. (1975). *Sociology of Social Criticism.* London: George Allen & Unwin Ltd.

Chakravarty, Meera (2003). "Communication for Women's Empowerment: New Dynamics and Priorities", In Kiran Prasad (Ed.) *Communication and Empowerment of Women: Strategies and Policy Insights from India* Vol.2, pp. 627–641. New Delhi: The Women Press.

Chakraborty, Sanghamitra (2004). "Mother India". *Outlook,* 1 November.

Crawford, C. S (1982). *The Evolution of Hindu Ethical Ideals*, Honolulu: The University Press of Hawaii.

Gaikwad, Rahi (2010). "Narmada Bachao Andolan Completes 25 years", *The Hindu*, 25 October 2010 retrieved from http://www.hindu.com/2010/10/25/stories/2010102556941500.htm

Heise, et al. (1994). "Violence Against Women: A Neglected Public Health Issue in Less Developed Countries". *Social Science and Medicine*, Vol. 39, No. 9, pp. 1165–1179.

Jebaraj, Priscilla (2010). "Nagoya Protocol a Big Victory for India, says Jairam Ramesh" retrieved 1 November 2010 from http://www.hindu.com/2010/11/01/stories/2010110153950300.htm

Jha, Suman & Nandini Oza (2006) "Dammed Emotions", *The Week*, 30 April. pp. 34–37.

Prasad, Kiran. (2006). "Cracking the Glass Ceiling: Rural Women Making News in India". *Media Asia*, Vol. 33, No. 3 & 4.

Prasad, Kiran. (2008). "Gender Sensitive Communications Policies for Women's Development: Issues and Challenges". In Katharine Sarikakis and Leslie Regan Shade (Eds.) *Minding the Gap: Feminist Interventions in International Communication*. USA: Rowman and Littlefield.

Sangvai, Sanjay (2000). *The River and Life: People's Struggle in the Narmada Valley*, Mumbai: Earth Care Books.

Sharma, V.D (1999). "Bishnois: An Ecoreligion". *The Hindu Survey of the Environment*, Chennai: The Hindu.

Shiva, Vandana (2009). "Food Meets Media", Opening Address, Media and Global Divides, IAMCR World Congress, Stockholm, 2008, *Nordicom Review*, Vol. 30, June 2009, pp. 11–31.

Swami Vivekananda (1985). *The Complete Works*, Vols 1–8, Calcutta: Advaita Ashrama.

The Hindu (2004, 28 March). "Q& A: Medha Patkar", retrieved 9 August 2007, from http://www.hinduonnet.com/2004/03/28/stories/2004032800971300.htm

The Hindu (2010, 25 October). "Sardar Sarovar: 40,000 Families Still to be Resettled" retrieved 25 October 2010 from http://www.hindu.com/2010/10/25/stories/2010102562341500.htm

United Nations Development Programme (2010), *Human Development Report 2010*, UNDP: Oxford/New York.

Vadakumchery, Johnson (1993). "The Earth Mother and the Indigenous People of India", *Journal of Dharma*, Vol. XVIII, No 1. pp. 85–97.

www.msn.com (2010) "Three Indian Rag Pickers to Address Climate Change Conference" retrieved on 6 October 2010 from http://green.in.msn.com/greennews/article.aspx?cp-documentid= 4442111 06/10/2010

Part II

Information and Communication Technologies for Development

5
Information and Communication Technologies for African Development: Proportional Technologies and an Ethics of Uses

Osée Kamga and Fabien Cishahayo

Introduction

How can the potential of Information and Communication Technologies (ICTs) be harnessed to reach the objectives of the Millennium Development Goals (MDGs)? This has been the key question in academic debates and various international forums. The debate is no longer focused on the issue of access, of bridging the digital divide, as was the case at the turn of the century. Although the technological divide remains in its various forms between developing and developed worlds, between Africa and the West, Western technologies have increasingly found their way into the "black continent". For instance, whereas Africa counted only 2 million mobile phone subscribers in 1998, the data released by the firm Informa Telecoms & Media indicate that the number reached 695 million in 2012.

Thanks to public and private initiatives, the support of international organizations such as the World Bank, the United Nations Educational, Scientific and Cultural Organization (UNESCO), the United Nations Development Programme (UNDP) and many others, as well as the backing of non-governmental organizations, the whole African continent is now connected to the Internet. In other words, today, each one of the 54 African countries has Internet connectivity. It is true that the high cost of communication charges and the low level of infrastructure cause Internet access to remain the privilege of few in most cases. Therefore, crucial challenges remain to be overcome to insure broader access in most African countries. Nevertheless, significant efforts have been deployed to improve Africa's access to ICTs. Also, there have been many attempts to articulate the ways in which these technologies can successfully respond to the social and economic challenges this continent faces. With such an objective, so far, the framework has been that of *appropriation*, that is, the processes through which we take foreign tools and use them in our own specific ways, in our own specific contexts, to reach our own specific objectives. Such a perspective has largely articulated the participative development paradigm, which emphasizes the grassroots' role in the development process; or, as Jan Servaes (2005: 6) puts it, "points to a strategy, not merely inclusive of, but largely emanating from, the traditional 'receivers' ". However, after analyzing the experiences of ICT use in Africa, we suggest that appropriation is no longer a pertinent framework, because it does not go beyond the notion that ICT users in Africa are mere receivers: it does not seize the whole spectrum of ICT experiences on the continent. Such a gap is due to a narrow conception that limits technology to its materiality.

Because of this, we argue that the broader framework of proportional technologies, which conceives technology as a practice, is more appropriate to account for African ICT experiences. We developed the concept of *proportional technologies* and have been playing with it in our small circle of young African scholars at the University of Sudbury, who are interested in the issue of ICT for development. This chapter is an attempt to test the concept outside our small group.

We begin the chapter by discussing various ICT phenomena in Africa. Then, we argue that the framework of appropriation is too limited to encompass such eclectic experiences. We believe that our position is far from a whim or a mere desire to suggest a new concept. With our belief that such a framework could fundamentally change ICT approaches to social change on the continent, the central question would no longer be: "How could a particular ICT contribute to the development of a given community?" Instead, it would be: "What technological ICT practices, in a given social context, can be profitably boosted?" Such an approach is key to sustainability, because it solves the issue of contextual adaptability that often accompanies innovation diffusion attempts. This is to suggest that the chances of sustainability being possible are higher because a practice is already part of the socio-cultural system. For instance, in a country like Cameroon, where soccer is the sport-king, initiatives to develop young talents or to improve soccer practice in the land have more chance of bearing lasting fruits than any attempt to introduce baseball or ice hockey.

In the end, we acknowledge that emerging ICT practices in Africa often challenge the regulatory framework, which makes it necessary to think of an ethics of uses.

ICTs in Africa: Some forms of insertion

Rural radio in Bankilare and the "domestication model"

A multimedia communitarian center was developed in Bankilare, one of the poorest villages in Niger, located some 240 km West of Niamey. This village of about 2,000 has no roads, no electricity, no telephones and no clean water. The experience was undertaken under the Rural Radio Network and Information Centers for Development (RURANET/ICD) program. It aimed to build a network of 160 self-managed, solar-powered, rural radio broadcasting stations across the country, with the ultimate goal of providing access to information communication for social and economic development. At the request of local women, the Bankilare center was established and backed by a UNDP Poverty Alleviation Programme, which considered Bankilare's communication problems a principal concern for women in the village. The center therefore came as a response to a need felt from within the community and expressed by its members. Equipped with broadcasting material powered by solar energy, it ran a self-managed Solar Rural Radio Unit (SEMRRU), with programming set up by local people focussing on local production in local languages. Sustainability was central to the project, the equipment used: transmitters, wind-up radios, and a WorldSpace digital satellite receiver, were all solar-powered. Bankilare's radio station started to operate in 2000.

This center, despite its apparent simplicity, is a technological exploit, given the poor economic, technological and infrastructural state of the village. However, the truth is that it is tailored for the village, both in terms of its operation and the purpose it serves. The broadcasters are given a small allowance of about $12 a month. The content of the broadcast is community oriented, that is, it disseminates indigenous knowledge, a "knowledge that people in a given community have developed over time, and continue to develop" (IIRR, 1966: 7).

We call this a model of "domestication" (Silverstone and Haddon, 1996), where a modern technology is harnessed to serve the basic concerns of traditional life. Weather, awareness of disasters or potential catastrophes, health and nutrition information, conflict resolution, entertainment, environmental conservation, as well as announcements from community members, are part of the broadcast subjects. The station reaches a radius of about 25 km. Therefore, "the farmer in the field; the herder out to pasture with his flock, and the woman doing laundry at the watering hole, all receive their fill of daily information without being restricted geographically to the village and its immediate surroundings" (Tchwenko, 2004). The accuracy of such information as weather is enhanced, enabling famers to plan their day and even their week accordingly. The announcement of potential or even actual catastrophes such as fires alerts the population to take the necessary precautions. The experience has been so successful that it has been replicated in many other villages all over the country.

Bus tour in Rwanda and the "Mutualisation of resources"

In Rwanda, the public transportation system has been given a substantial make-over. Since 2009, two buses equipped with state-of-the-art technology worth $200,000 each and financed by the International Development Association have been touring the country. These mobile tele-centers, equipped with a generator and a server of 22 computers, each with Internet access, allow farmers, small- and medium-sized enterprise (SME), and associations of young people, to access the digital world. The buses are equipped with printers, photocopiers, a scanner and other multimedia facilities. The communities they visit are provided with information related to agricultural commodity market prices, business opportunity information, health, best agricultural practices and seasonal weather forecasts, all of which are of great importance to rural farmers, traders, women, students and entrepreneurs. Such initiatives have been highly useful in a country where more than 70 per cent of the private sector is composed of small and medium enterprises (SMEs), most of which operate in rural areas and do not individually have the means to afford such equipment.

Mobile phone and innovative usages

In Côte d'Ivoire, the cellular industry developed largely thanks to the phenomenon of "mobile booths", which, although being marginal to the formal economy, enjoy great popularity. Because of the legal status of these "booths" or, more precisely, as they do not have one, the mobile industry in Côte d'Ivoire presents a hybrid figure, extremely productive as an analytical phenomenon. What characterizes the "booths" is, on the one hand, their completely anarchistic space distribution and, on the other hand, the total absence of formal rules. The phenomenon has developed at the margin of the regulatory system and has spread throughout the country. Its actors in the field are more or less unaware of the role or even the existence of the regulatory agency, the Agency of Telecommunications of Côte d'Ivoire (ATCI). A "booth" structure is very elementary. The unit consists of an individual, often under a parasol, sitting with a mobile telephone in front of a small removable table on which rests a sign indicating the tariffs. This is the traditional configuration of the mobile "booth", although one can observe certain alternatives to this in terms of strategy.

Since the research we conducted in 2002/2003, the mobile booth phenomenon in Côte d'Ivoire has evolved. Besides retailing the communication traffic, booth owners are now transferring money with mobile phones, without having to appeal to any financial institutions; the whole transaction takes place outside of it. The concept is based on the use of airtime as a monetary value. The sender pays a booth operator, who in turn contacts another operator at the location where money is to be sent. When the second operator agrees, his booth account is electronically credited with the equivalent amount he is supposed to hand over to the receiver. This type of operation works especially well with booth owners who have enough cash in hand to meet customers' needs. Fees paid in these transactions are on average three times lower than they would have cost with financial institutions such as Western Union or other money-transfer agencies. Clearly, it is a practice that consists of "poaching" (de Certeau, 1984; Frow, 1991), using the mobile phone industry.

For de Certeau, any social system combines upstream "strategies" and downstream "tactics". He suggests that the rationalized, dominant structure of the system has to cope with another one characterized by its tricks, its opportunism and its poaching, without its own commodities, but with the skills to make use of those imposed upon it. Here, the booth owners do not have anything like a bank; they do not own the networks they use for their transactions, nor are they legally authorized to perform such transactions. They simply trick a system in which they have no power. They "poach" to meet the challenges they face. The mobile booth deployment moved mobile operators to readjust their various offers. With the mobile booth now being used for money transfers, if mobile operators want their share this potential manna, they would need to make new and more fundamental adjustments, which would entail the technological dimension – moving toward smart phones as in

Indonesia – and the regulatory dimension, compliance with laws that regulate the financial sector.

Illiteracy and the delegation of usage

One of the common practices in Africa is what has come to be termed the "delegation of usage"; it intervenes in commercial transactions and in interpersonal relations. For instance, illiterate Senegalese and Malian economic operators entrust their searches for information about commercial opportunities to a third party – an investor who has banked on this business of mediation. Thus, they avoid physically making a trip to the country to be investigated, as were their previous habits. Once the investigations are completed, the orders are placed and payments are usually made through some traditional mode of money transfer such as Western Union.

With regard to interpersonal relations, the delegation of usage takes the shape of help provided by cybercafé employees to illiterate customers, or those unable to handle hard or software. For example, in the case of email, customers can utter their message or provide a handwritten message that the employee takes care to type and to send. The customer will come back later to inquire about the answer. This mediation uses the mode of payment per message instead of per connection time.

This delegation of use has another advantage for the end user, especially when the formula is pay per minute. It is worth noting that a user unfamiliar with computers will spend more time than necessary to complete a task. Also, since most of the costumers use free e-mail addresses – gmail, yahoo, hotmail and so on, the inherent popup ads on these sites contribute to slowing down operations, which ultimately becomes costly for users. Therefore, paying a cybercafé agent appears to be a simple and more efficient solution for an inexperienced or a computer-illiterate user. This form of usage creates specific "conventions" (Boltanski and Thévenot, 1993) for use in societies such as in Africa, where illiteracy and verbalism prevail. Such delegation of usage is not new in rural Africa where the public writer has always existed, especially in areas where the literacy rate is low. In the end, the delegation of usage is not likely to boost computer literacy on the continent: the costumer who delegates a usage to a third party makes use of the computer service, but is not a computer user in the hard sense of the term, that is a human interacting with the machine.

The limits of the appropriation framework

Overall, appropriation theories suggest that social changes depend on people's ability to adopt ICTs and integrate them into their way of life. Around this basic conception, a number of theoretical perspectives have developed during the last 50 years, each one suggesting how the adoption process can succeed. The various approaches can be classified into two categories: the prescriptive and the adaptive approaches.

By prescriptive approaches, we refer to theoretical perspectives that consider technology as a pill and therefore focus on finding the best way to make receivers swallow it. This was the case with the modernizing theorists, as exemplified by Rogers' sequential innovation, diffusion theory, which suggested that a successful diffusion of innovation proceeds in five stages: knowledge, persuasion, decision, implementation and confirmation (Rogers, 1962). However, the works of Colin Fraser and Sonia Restrepo-Estrada (1998), who pioneered the use of communication techniques for development endeavors in Latin America while emphasizing the importance of communication, demonstrated that: (1) its one-way-flow communication does not work; and (2) targeted populations have to be integrated in the process of their own development. Also, this prescriptive technology is essentially an artefact, a tool or device which, integrated in a given context, can bring changes. However, as Philippe Breton (2000) observes, the demiurgic conception of technology has been demystified in many works in the field of the sociology of techniques. They have shown that the socio-economic and cultural components are essential in the adoption process of a technology. Madeleine Akrich (1993) articulated this view in terms of the user's "frames of action". Even if the designer anticipated the frame in which the user will exploit his/her device, the environment of the usage can present specific obstacles that the user has to contend with. Flichy (1995: 139) speaks of the adjustment of the user to the technical object and suggests that it goes beyond technical performances. Overall, these works suggest that what matters is not the technology per se, but the uses people make of it, and those uses depend on a whole range of social, economic and intellectual resources the users bring to the table, as well as on the existing power struggle in a given situation. This explains why, in some particular contexts, people "boycott" innovative technologies since they consider the new power play these innovations introduce to be a threat to their social position, or an obstacle to the scope of their control over their organization (See Moss-Kanter, 2001).

The adoptive approach, on the other hand, refers to a theoretical perspective that rests on the assumption that technology can do works even its makers did not imagine. In this perspective, ICT users in the South are perceived as people with inventiveness and ingenuity, capable of using the technologies in authentic ways, beyond the standards set by dominant structures. The theoretical framework for such a view is provided by Michel de Certeau's notion of "usage" (1984), which suggests that there is always a variation between projects defined upstream and the smartness of operating downstream. Boulou de B'béri and Margaret Cormier capitalize on de Certeau's views with the notion of "*marronage*" – a signifier referring to slaves' paradoxical resistance and attraction to colonial power in Caribbean studies (de B'béri, 2008: 200); to account for the uses of the mobile phone in Haiti. Here, the attraction to the mobile phone is evidenced by the penetration rate; and the resistance is seen in the tacit refusal to subscribe to modes of utilization prevailing in the developed world. Cormier and de B'béri therefore conclude that "the integration of the mobile telephone in the Haitian socio-cultural, economic and political context offers users the opportunity to transform pre-established cultural practices, to create new ones, and reveals the capacity of layouts unique to them"[1] (Cormier and de B'béri, 2010: 195). The practice of beeping in Africa are definitely a form of "*marronage*". In this practice, to avoid paying the communication cost or to let their wealthier correspondents pay for it, mobile phone users in African have invented a series of codes in their communication exchanges. For instance, in many places, one beep often means: "I will be late"; whereas two mean: "the appointment is canceled" and so on. Haevens (2007) has made a good description of the phenomenon. However, the concrete implications of such findings for development remain to be articulated.

It also appears that this adoptive approach reduces technology to its material dimension, even if the users' ingenuity is here recognized. Because of such a limitation, this approach is unable to account for experiences where technologies, though factors in the practice taking place, are not specifically used. The study conducted by Christian Agbobli (2008) on cybercafés in Senegal provides an interesting illustration. The centers aimed at helping Senegalese people to take advantage of the benefactions of the digital era. However, Agbobli found that most of the cybercafés' visitors were not there to use computers, but simply to engage in social and political discussions.[2] The business place appeared as an *agora* of some sort, where Senegalese met daily to discuss a whole range of issues, from politics to sports, business and other social considerations. In other words, the presence of the new technology did not introduce the change hoped for. The expectation that such centers would serve primarily for Internet research and other computer-related business was not met. From a strictly appropriation standpoint, this might be considered a failure. However, as Agbobli suggests, the notion of appropriation fits if it is broadened:

> Analysis of the telecentres shows that the interactive situation we find there demonstrates Internet use from a perspective larger than that of the technical tool itself. We find this in the *lato sensu* dimension of Internet use. In fact, for the managers as much as for the users, the Internet is a signifier much larger than the content of a computer. We could perhaps posit that Internet use can be understood to be the appropriation of a centre as a place for communication. If the goal of the Internet is to connect computers around the work with one another, in the case of Acacia's telecentres, Internet use manifests itself according to broader social phenomena depending on where access points are situated. Consequently, Internet use consists of an appropriation of the physical space in which it is located. It is as if the active sphere of the Internet has grown to include the actual telecentre. However, this appropriation corresponds to the contextual reality of use demonstrated by the fact that traditional communication, based on orality and face-to-face communication, is present in Senegal, even in the context of Internet use.
>
> (Agbobli, 2008: 24)

We suggest that, instead of broadening the framework of appropriation, a new framework should be adopted, that of proportional technologies. Rose Tchwenko (2004) appropriately urges us to "discard any preconceived ideas about the needs of people in the developing world".

The framework of proportional technologies

The above examples show that a variety of phenomena take place around ICT artefacts in Africa:

- New practices can emerge. This is certainly the case with the mobile booth phenomenon in Côte d'Ivoire and other forms of mutualization of uses in Africa.
- In some cases, a technological device simply enhances existing communication practices. For instance, a Bankilare farmer now has access to more accurate weather forecasts. Knowing whether or not the

following day will be sunny becomes extremely important, affecting the types of activities undertaken the next day. For example, the farmer might know whether to spread his corn out to dry it or keep it inside before he leaves the house in the morning. It is worth stressing that what is broadcast serves to improve existing community practices. In other words, the goal is not to change what these farming populations do, but to help them become more efficient in what they do.

- The technological set-up might simply serve as a pretext for other activities without being in itself a factor in those activities, as Agbobli's works have demonstrated. In other words, what really matters is *context*.

First of all, the *social* context: we here refer especially to social structures, the forms of social bonds, or the social organization. Is it articulated by reliable institutions or by an arbitrary power structure? Are the laws enforced or are they generally loose in their application? Contrary to Harold Innis (1964) or Marshall McLuhan (1964), we are not convinced that the introduction of new technologies in a society determines the way it is organized, or the perception its members have of the world surrounding them. Instead, the form of social structure shapes the way technological practices take place or expand. This is, for instance, the case with the development of the mobile booth phenomenon in Côte d'Ivoire, a practice that emerged at the margin of legality and expanded because of slackness on the part of the Agence de Télécommunications de Côte d'Ivoire (ATCI) in enforcing the law. Social structure has not changed in the country and the booth practices are now moving forward with financial services.

Second, there is the *economic* context: both the economic level and the economic structure. The mobile booth practice developed in Côte d'Ivoire as part of the flourishing informal structure of the economy, whereas in Bankilare the implementation of the radio had to take into account the economic level of the population. This meant the use of solar power instead of other sources that would make the maintenance of the structure unsustainable; it meant providing incentives to broadcasters according to the limited resources of the community; it also meant solar and manual-powered FM receivers to help the population avoid expensive battery renewal. In sum, cheap equipment that does not necessitate expensive maintenance was privileged.

Third, there is the *cultural* context: by culture, we understand the way people give meaning to the world and the things around them, the "shared conceptual map" (Hall & du Gay, 1996). What does a computer mean for the community? How do the community members see a mobile phone? A radio set? The answer to these questions has a bearing on whether and how these artefacts would fare in a given context. Without a basic common understanding of a technological device and its usefulness within a certain community, its deployment would never take root. The fact is that social forces have to converge at some level for an artefact, whether it be a communication artefact or not, to become established. The artefact has to be used in an incremental instead of a disruptive way for accepted social practices. Kadi Kohler (2008: 20) notes, for instance, that "In a strongly Islamic country such as Niger, speaking of women's liberation even on the air can present a challenge and even lead to the closing down of a radio station. Radio stations that give women positions of responsibility are rare." Therefore, if radio is perceived primarily by power structures as a threat to a way of life or to social hierarchy, its sustainability might be jeopardized.

We suggest that an artifact has a meaningful life only within a social, economic and cultural context and therefore cannot be singled out as the ultimate factor in social change. Its mere presence in a given context is meaningless. What really matters is whether and how it will become a sustainable component of a social practice. This sustainability depends on a whole range of social, economic, and cultural factors. Therefore, we privilege the conception of technology as a practice that involves "organization, procedures, symbols, new words, equations, and, most of all, a mindset" (Franklin, 1990: 12).

The framework of proportional technologies lies in the assumption that it is through practices that social changes occur. The question should therefore be: what emerging practices bring positive changes to the society? To the community? We adopted the term "proportional" for its mathematical meaning, which suggests both variation and constancy. "When two quantities vary in such a way that one of them is a constant multiple of the other, the two quantities are 'proportional,'" note Stanley et al. (2003: 2). Therefore, the concept of proportional technologies entails the recognition that a technological practice may change trajectory in an evolving social system, as well as the suggestion that both (the practice and the system) have to remain in constant adequacy for tangible results to be sustainably produced.

With such a framework, temporality becomes a key element in the elaboration of development projects. It takes time to observe technological practices in order to determine both their social usefulness and their sustainability. Therefore, social actors, who are the prime observers of their own reality, are in a better position to identify their needs and the type of projects that work. When Bankilare's women expressed their desire to have a local radio, they were already familiar with the radio, since their community had received, in August 1998, wind-up radio units from the African Center of Meteorological Applications for Development (ACMAD) which gave them access to information from outside their region. From their observation and radio experience, they could envision the results of a local station for their community. In other words, the potential of a local radio broadcast was not an abstract concept to them. Also, when the UNDP helped Ivorians who fled the civil war ravaging the northern part of the country and found refuge in Abidjan to start a mobile booth business, it was because they observed the effectiveness of the practice. Tens of thousands of people were already making their living with such a business.

By proportional technologies, we mean fitting practices, born within a given context and consistent with other established useful practices of the context. If such a framework can be situated within the paradigm of a participatory approach to development, it also takes into account the position stressed by Bourdieu (1980), according to which social practices have intrinsic logic, independent of preconceived purposes, and motivated primarily by practical ends. In other words, we cannot expect the trajectories of a given technological practice to be the same from one society to another, from one community to another, even if these communities present similar characteristics. Therefore, the reproduction or transposition of a successful experience in development projects carries the risk of disappointment. We insist on the organic character of communicational practices, that is, their emergence as intrinsic constituent of the social fabric, as a more secure way to sustainability.

Setting an ethics of uses

In the dynamics that articulate ICT penetration in Africa, the point of tension is often between the "legal" and the "moral", between what the law authorizes and what the common right allows. The uncoordinated deployment of emerging practices to a regulatory framework can really slow down the benefits of ICT penetration in Africa. The example of the mobile booth in Côte d'Ivoire illustrates how emerging ICT practices often take place outside of any legal consideration. This space of liberty has contributed to the expansion of the phenomenon, but it may well become a hurdle. Anyone anywhere can be a booth owner, but a public authority can, at its discretion, forbid the practice in its jurisdiction, depriving tens of thousands of customers of their main means of communication and making it harder for hundreds of booth owners to earn their living. In other words, without legal protection of some sort, the practice appears highly precarious. Yet, a new service, the money transfer, is progressively gaining ground. It is not necessary to mention that money transfer pertains to another sector of activities, under other sets of regulations, which are totally ignored by the booth operators.

Although such a system provides needed services to individuals and appears lucrative to operators, it is hard to imagine its sustainability in the long run. Anarchy and precariousness, in our view, constitute a good recipe for chaos. Such a situation testifies to a lack of cooperation that exists between key actors of the ICT system in Africa, namely the State, the industry and the public. On the one hand, the users' inventiveness is often disregarded by the State and considered as some sort of deviance which, at best, should be ignored and, at worst, fought against. On the other hand, the industrialists often disregard users' law infringement practices, as long as these practices generate maximum revenues. Such an environment, based on arbitrary power, gives rise to abuses. In Côte d'Ivoire, for instance, newspapers have criticised mobile phone operators for "exploiting" booths' owners. There is therefore a need for cooperation between actors, namely communities, government and private operators. There is also a need for elaborating a code of ethics aimed at articulating the interests of the booth owners, the operators and the State around the transcendent value of social progress.

How thus to establish principles which mark out everyone's responsibilities and prevent abuses for the common good? It is here that the Habermasian approach of communicative ethics can be useful. Such an approach suggests that issues should be publicly debated, with the participation of the various stakeholders. Habermas wrote:

> Therefore, the protagonists, even in situation [sic] of conflict, have to continue their interaction with a communicative activity attitude. From a participant starting point in the first person, each one has to extend himself to the other as the second person, with the intention of understanding him, instead of treating him as an object, and thus, instrumentalizing him for personal purposes.
>
> (Habermas, 1998: 85)[3]

The social dialog, vehicle of mutual understanding becomes the means for individual and collective self-fulfillment, because it acknowledges the recognition of a multiplicity of interests, and the acceptance that there is a fundamental shared value, namely, social progress. It is about defining conditions, based on the principle of fairness, which would solve tension between the legal and the moral, conditions to which everyone would subscribe to make the system work. Such an ethics would clean up the economic environment, which is always essential to attract investors and would provide emerging practices with the official recognition needed to thrive harmoniously in a social system.

Conclusion

The emerging ICT practices discussed in this chapter call for another way to look at technologies for development in Africa. They suggest that we should stop thinking of them in terms of imported artifacts carrying some sort of intrinsic value or potential. Such a perspective has dominated development strategies for a long time, with the result that many useless infrastructural systems, technological or otherwise, (the "white elephants" Fottorino [1992] speaks of) sprinkle the continent. Therefore, the conversation should not be about the ways ICTs can help. Such preoccupations appear detached, separated from the context, with the dubious underlying assumption that there exists an inherent way for ICT to foster development.

This way of thinking does not take into account the fact that social practices do not emerge and deploy according to a deterministic logic. More specifically, the question of how mobile phones, for instance, can contribute to development might simply give way to hypothetical answers. The question of how to use booth practices to help a social category in Abidjan, however, will lead to more concrete answers, because the practice is already part of the social system. This is precisely the approach the UNDP used to help refugees in Abidjan early in this century during the Ivorian civil war, with a great measure of success (Kamga, 2005, 2006).

Introduced in a given environment, the life of a technology or the direction it will take is not guaranteed. The context can adopt it or reject it altogether. Within the appropriation framework, a rejection is commonly explained by the notion that technologies conceived primarily for Western societies embody Western ideologies, a Western vision of the world or Western culture, which makes the articulation with Africans difficult Misse Misse (2004). We reject this notion, given the fact that other societies with anchored cultural traditions have clearly demonstrated, through their development experiences, that they can thrive with modern world artifacts.

One of the objectives of the MDG consists of mobilizing ICTs to undertake major challenges of development such as ending extreme poverty and hunger. For such a goal, emerging practices in Africa should not be disqualified on the grounds that they do not correspond to experts' prescriptions or meet preconceived expectations. The examples considered in this chapter show how useful practices often emerge from regular people and generally outside of standards set by the State and other structures of power. Occasionally, these practices appear at odds with the regulatory framework and this paradox (being socially useful, but at the same time being lawfully unsocial) creates a precarious system.

This is where we argue for an ethics of uses, the goal of such an ethics being to economically encourage emerging practices while providing them with a statutory stand. This should be done without locking such practices into hard legislative rules, for such rules run the risk of creating new margins where the social actors involved would seek refuge. The first effective step would certainly be that of a mandatory state engagement with stakeholders. The Right to Information Act adopted by the Indian government can be inspiring. It "empowers common people to raise questions with government directly and the concerned authority has to reply back within 30 days with proper explanation" (Bhardwaj, Sharma, Kumar, 2013: 168). Such engagement, which Habermas (1998) calls "communicative activity attitude", opens the way to mutual understanding among stakeholders and creates the conditions for consensual decisions. In the end, a flexible approach, or a form of regulation that does not choke emerging ICT practices, creates a minimum level of predictability and ensures their sustainability. In other words, innovative practices should stimulate statutory innovations with enough flexibility to accompany the creative run-up of social actors.

Notes

1. Free translation of: "l'intégration de la téléphonie mobile dans le contexte socioculturel, économique et politique haïtien offre aux utilisateurs l'occasion de transformer des pratiques culturelles préétablies, d'en créer de nouvelles, mais aussi de révéler des capacités d'agencements qui leur sont propres."
2. A similar study has been conducted by Claire Scopi with immigrants in Paris.
3. Free translation of: "Les protagonistes doivent donc, même en situation conflictuelle, poursuivre leur interaction dans une attitude d'activité communicationnelle. Chacun doit, à partir de la perspective du participant à la première personne, se rapporter à l'autre en tant que seconde personne dans l'intention de s'entendre avec lui plutôt que de le traiter en objet et de l'instrumentaliser à des fins personnelles."

Bibliography

Agbobli, Christian (2008) "Internet and development in Senegal: Towards new forms of use". *Africa Media Review,* 16 (2), 11–28.

Akrich, Madeleine (1993) "Les objets techniques et leurs utilisateurs, de la conception à l'action", *Raisons pratiques*, n°4 *Les objets dans l'action*, pp. 35–57.

B'béri (De), B. E. (2008). Africanicity in Black Cinema: A Conjunctural Ground for New Expressions of Identity. In Cultural Studies in Africa/African Cultural Studies. Cultural Studies, London: Routledge. Vol. 22 (2), pp. 188–209.

Bhardwaj, A., Sharma, M., & Kumar, V. (2013) "Identity management for social development in India" in Viskas Kumar and Saurabh Mittal (Eds.), *Leveraging Information Technology for Competitive Advantage*, Proceeding of the International Conference on Leveraging Information Technology for Competitive Advantage, Feb. 21–22, 2013.

Boltanski, L. &Thévenot, L. (1993). A qui convient la théorie des conventions? *Réseaux*, 62, Nov–Dec, 137–142.

Bourdieu, P. (1980). *Le sens pratique*, Paris, Gallimard.

Breton, Philippe (2000). *Le culte de l'Internet*, Paris, La découverte.

Cishahayo, Fabien (2010). *Communication, développement et appropriation des nouvelles technologies de l'information et de la communication en Afrique subsaharienne francophone. Jalons pour une approche critique*. Thèse de doctorat. Montréal, Université de Montréal.

Cormier, M., & de B'Béri, B. E. (2010). "Usage et rappropriation contextuelle du téléphonemobile à Port-Au-Prince". In de B'BÉRI, B. E. (Ed) *Les Cultural Studies dans les mondes francophones*. Ottawa, Presse de l'Université d'Ottawa. pp. 169–199.

Certeau, M. de (1984). *The Practice of Everyday Life* (trans. Steven Rendall), Berkeley: University of California Press.

Escobar, Arturo (1994). *Encountering Development The Making and Unmaking of the Third World*, New Jersey, Princeton University Press.

Flichy, Patrice (1995). *L'innovation technique. Récents développements en sciences sociales. Vers une nouvelle théorie de l'innovation*, Paris: La Découverte.

Fottorino, Eric (dir.). (1992). *Besoins d'Afrique*, Paris, Fayard.

Fraser, Colin and Restrepo-Estrada, Sonia (1998). *Communicating for Development Human Change for Survival*, London, I. B. Tauris.

Franklin, Ursula (1990). *The Real World of Technology*, Concord, House of Anansi Press Limited.

Frow, John (1991). "Michel de Certeau and the practice of representation", *Cultural Studies*, 5:1, 52–60.

Habermas, Jürgen (1998). *Le discours philosophique de la modernité*, Paris, Gallimard.

Hall, Stuart & Paul du Gay (eds.) (1996). *Questions of Cultural Identity*, London, Sage.

Haevens, A. (2007). "Phone Credit Low? Africans go Beeping". 26th September, 2007, http://www.yahoo.com

IIRR. (1966). *Recording and using Indigenous Knowledge: A Manual. Silang, Cavite, Philippines: REPPIKA*, International Institute of Rural Reconstruction.

Innis, Harold (1964). *The Bias of Communication*, Toronto, University of Toronto Press.

Jennings, Ray (2000). "Participatory Development as New Paradigm: The Transition of Development Professionalism", document Prepared for the "Community Based Reintegration and Rehabilitation in Post-Conflict Settings" Conference Washington, DC October 2000, http://www.usaid.gov/our_work/crosscutting_programs/transition_initiatives/pubs/ptdv1000.pdf

Kamga, Osée (2005). *De l'utopie du développement à l'analyse des pratiques* communicationnelles: les usages de la téléphonie mobile en Côte d'Ivoire dans *une perspective de praxis africaine,* Montréal, UQAM.

Kamga, Osée (2006). "Mobile phone in Côte d'Ivoire: Uses and Self-fulfillment", *Proceedings of the International Conference on Information and Communication Technologies and Development,* (ICTD, 2006) Berkeley, CA.

Kohler, Kadi Souley B. (2008). Community Radio in Niger. In *Women's Empowerment and Good Governance Through Community Radio. Best Experiences for an Action Research Process,* AMARC.

Lerner, Daniel (1958). *The Passing of Traditional Society,* New York, The Free Press.

Misse Misse (2004). L'aporie de la communication sociale pour le développement. http://www.ugrenoble3.fr/chaire_unesco/Textes/misse/seminaire5.htm

McLuhan, Marshall (1964). *Understanding Media: The Extensions of Man,* NewYork, McGraw-Hill.

Montreiro, E. (1998). Living with Technology. *Scandinavian Journal of Information Systems,* 10(1&2), 249–254.

Moss-Kanter, R. (2001) *E-Volve: Succeeding in The Digital Culture of Tomorrow.* Boston, Harvard Business School Press.

Rogers, Everett M. (1962). *Diffusion of Innovations.* Glencoe, Free Press.

Rogers, Everett M. (Ed) (1976). *Communication and Development Critical Perspectives.* London, SAGE Publications.

Scopi, Claire-Langlet (2004). *Représentations des TIC en milieu migrant: le cas des « boutiques de communication » de Château-Rouge.* Doctorate thesis, Nanterre, Université Paris 10.

Servaes, Jan (2005). Mapping the new Field of Communication for Development and Social Change, Paper presented to the Social Change in the 21st Century Conference Center for Social Change Research, Queensland University of Technology, 28 October 2005.

Silverstone, R., & Haddon, L. (1996). Design and the Domestication of Information and Communication Technologies: Technical Change and Everyday Life. In R. Mansell & R. Silverstone (Eds.), *Communication by Design the Politics of Information and Communication Technologies,* Oxford UK, Oxford University Press.

Stanley, Dick, McGowan, Diane, & Hull, Susan Hudson (2003) Proportionality. http://www.utdanacenter.org/mathtoolkit/downloads/support/proportionality. pdf

Tchwenko, Rose (2004). *The Bankilare Experience: An Example of a Successful Collaborative Effort to Bridge the Digital Divide.* http://www.cto-ict.org/index.php?dir=04&sd=30&aid=1055

Thussu, Daya Kishan (2000). *International Communication Continuity and Change,* London, Arnold.

UNDP (1996). *Human Development Report,* New York, UNDP.

UNDP (2005). Sub-Saharan Africa – the human costs of the 2015 "business-as-usual" scenario. *Human Development Report Office.* http://content.undp.org/go/cms-service/stream/asset/?asset_id=1397338

6
A Role for Universities in ICT for Development Interventions

Royal D. Colle and Tran Van Dien

Two important movements have surfaced during the past decade that have the potential to influence social change initiatives in developing nations. The first is the growing importance of Information and Communication Technologies (ICTs), fueled especially by the emergence of more affordable computers, mobile phone devices, and broadband Internet. Their potential lies in the recognition by development agencies, including both government and non-government organizations, that ICTs can play a major role in reaching targets such as those in the year 2015 (MDGs). A major challenge is to bring these digital resources into the lives of those who can benefit most from them.

A second movement is the attempt to *reinvent* higher education institutions to make them more relevant to the world around them, and to include them among the partners in development. This chapter brings these movements together in exploring the role that universities can play in nurturing the establishment of community access centers (or tele-centers) – places where disadvantaged people can join the information society and improve their well-being.

Innovative ways of communicating tempt us frequently. Now, early in the twenty-first century, there are about 65 million tweets traveling through space every day, or 750 *per second*. This is impressive communication. However, a 2009 study by the United States market research firm Pear Analytics suggested that more than 40 per cent of twitter messages are "pointless babble".[1] We could probably say the same about the content of television programs, Internet blogs, and some magazines and newspapers. This raises the larger question of whether we are getting appropriate benefits from our ICTs. We explore this issue by examining two parallel streams related to communication for social and behavioral change, focusing especially on developing nations. One is the emergence during the past decade of the concept and practice of Information and Communication Technology for Development (ICTD). The other is a parallel movement to reinvent universities and other higher education institutions so that they contribute more significantly to development issues. We will explore opportunities to join these forces – with universities helping sustain ICTD institutions and building ICTD human resources. The principal theme of this chapter is that combining these streams should have high priority in strategies for reaching important socio-economic targets such as the MDGs.

The emergence of ICTD

For many years in the last half of the twentieth century, many of us struggled to get national governments and development agencies to incorporate communication more prominently as a partner in development programs. Erskine Childers worked tirelessly on getting United Nations agencies to recognize the role that communication could play in health, agriculture and other interventions.[2] Colin Fraser did much the same at the Food and Agriculture Organization. Writing in 1998, Fraser noted that although communication "has clearly shown its usefulness and impact in change and development actions, its role is still not understood and appreciated to the point that it is routinely included in development planning".[3] The struggle changed in the 1990s as computers and the digital world became more prominent forces in industrial countries. Information and communication technologies became increasingly important in the everyday lives of people across the globe. Many international gatherings awoke to the potential. Here are just a few examples. A decade ago, the eight major industrial nations (the G8) acknowledged that information and communication technology "is one of the most potent forces in shaping the twenty-first century [and] its revolutionary impact affects the way people live, learn and work, and the way government interacts with civil society".[4] Emerging from the G8 meetings in Japan was the Okinawa Charter on the Global Information Society. Its framers announced that "this Charter represents a call to all, in both the public and private sectors, to bridge the international information and knowledge divide". The Charter also renewed a commitment of the G8 nations "to the principle of inclusion: everyone everywhere should be enabled to participate in and no one should be excluded from the benefits of the global information society". The G8 launched a major effort to strengthen all nations' potential to be part of this information age by starting with a Digital Opportunity Task

Force (DOT). The DOT noted the relationship between high-priority international development goals (such as the MDGs) and communication, and emphasized that harnessing the power of information and communication technologies could contribute substantially to realizing *every one* of these goals – either directly (e.g., through greater availability of health and reproductive information, training of medical personnel and teachers, giving opportunity and voice to women, expanding access to education and training), or indirectly (through creating new economic opportunities that lift individuals, communities and nations out of poverty).

The two-part World Summit on the Information Society held first in Geneva in 2003 and then in Tunis in 2005, underscored the role of ICTs in promoting social change. It presented a vision of Connecting the World by providing digital connectivity to 800,000 villages by 2015. The Summit also reaffirmed the potential of communication and information in promoting the MDGs, and emphasized the importance of ICTs for political, social and economic development.

Since the G8 meetings, the Summits and a host of documents, books and reports, we have seen the emergence of the eWorld and them World: eGovernment, eHealth, eCommerce, eEducation, mAgriculture and others. The inescapable message in all this is the importance of bringing disadvantaged and marginalized people across the "digital divide", or as the G8 said almost a decade ago: "establish the principle of inclusion: everyone everywhere should be enabled to participate in and no one should be excluded from the benefits of the global information society."

However, some data on the new information technologies demonstrate a gap between countries – and between populations within countries – as to the availability and *use* of digital resources. Ownership of computers and access to the Internet are important relevant issues. There are credible estimates that people in the higher income countries are about 22 times more likely to use the Internet than persons in lower income nations. According to International Telecommunications Union estimates, by the end of 2010, 71 per cent of the population in developed countries were online compared with 21 per cent of people in developing countries. Figure 6.1 illustrates the point.

The figure shows that the highest penetration of the Internet is in North America, with about 77 per cent of the population being connected, while in comparison, 11 per cent of Africans are connected. Telecommunications specialists look to their experience with telephone service for terminology to apply to the Internet situation. They use the term "universal service" to describe the *one-telephone-to-one-household*

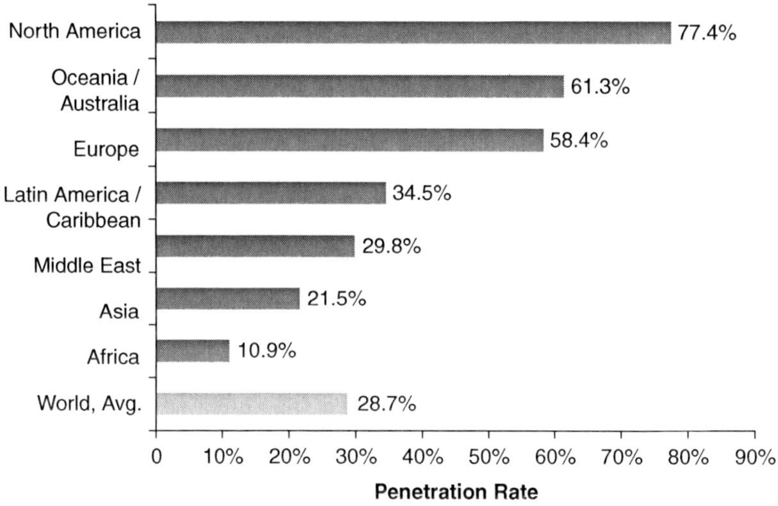

Figure 6.1 World Internet penetration rates by geographic regions – 2010
Source: Internet World Stats – www.internetworldststs.com/stats.fmt. Penetration Rates are based on a world population of 6,845,609,960 and 1,966,514,816 estimated Internet users on June 30, 2010. Copyright ©2010, Miniwatts Marketing Group.

ownership pattern. The more viable strategy for developing countries (before cell phones) had been "universal *access*" – meaning that a telephone should be within a reasonable distance for everyone. Telephones would be a *shared* resource. The sharing took place through such devices as public telephone booths, commercial call centers and "telephone women".[5] Turning to newer technologies, the International Telecommunications Union (ITU) estimates that in developing countries, only 13.5 per cent of people have the Internet at home. Thus, in regard to computers and Internet access, an interim target for dealing with the digital divide is this concept of universal access – to have computers and network coverage shared in public places and accessible to all. Thus we see the widespread prevalence, particularly in urban communities, of cybercafés and some four million Wi-Fi "hot spots".

In more rural areas, tele-centers have become one answer to the ICT inclusion-and-access challenge. They tend to be more focused on serving the community as compared to making a financial profit. Telecenters originated in the mid-1980s in Scandinavian countries as an effort to help people in rural areas become part of the information economy, particularly by making modern office equipment available to them on a *shared* basis. Although the idea spread through much of the industrialized countries for the next decade, it was the emergence of moderately priced computers and digital networks in the 1990s that launched a newer brand of these public access centers – with a variety of names such as community technology centers, village information centers, telecottages, and community learning centers. The United Nations Educational, Scientific and Cultural Organization (UNESCO), the World Bank, the ITU, the UN's Food and Agriculture Organization (FAO) and overseas development agencies such as those of Canada, Switzerland, the United Kingdom and the United States gave the tele-center movement significant support and propulsion as they perceived a link between ICTs and the social change that is one of the keys to sustainable development.

Challenges to the ICT-for-all strategy

Despite the enthusiasm and the support given to the shared access concept by international and national governments and non-governmental organizations, significant challenges impede more widespread scaling up of development-oriented tele-centers. Foremost among these is the lingering question concerning the relative benefit of putting money into information technologies versus the demands for other kinds of development resources such as roads, water supplies and health clinics. This stems, in part, from the perceived lack of convincing data about the impact of the Internet – a skepticism captured in a book titled *Overselling the Web? Development and the Internet* by World Bank expert Charles Kenny.[6] He points out that "As of December 2005 ... there had not been survey-based, academically rigorous study of the economic impact of an Internet access program in any developing country." While there is no evidence of a cause-effect relationship, the question of relative return on investment may help explain the minor attention given to information technologies in the Poverty Reduction Strategy Papers, the policy papers drawn up by developing nations as planning documents for encouraging broad-based growth and easing poverty.

Because access to the Internet is one of the principal attractions of tele-centers, the quality of service provided by the Internet influences the potential benefits of a tele-center. Availability and cost of broadband connectivity, which affects downloading speed as well as the delivery of more complex (often graphically enhanced) web pages, is problematic for many tele-centers, particularly those in Africa. Wireless connectivity has become one of the solutions, especially for accessing remote and hard-to-reach locations, although cost–benefit considerations continue to be an issue. Also, the characteristics of Internet content itself relate to consumer desire for tele-center services; for example, the content's relevance to local needs, the language of data-based material and websites, and the style of presentation are among the factors that tele-centers need to address to create a community-oriented, *demand-driven* tele-center.

Another challenge that tele-centers face is their sustainability. Tele-centers obtain operating funds from service fees (e.g., for sending email or providing computer training), from public subsidies and from contributions (including volunteer help). Most begin with donor funds and then struggle for sustainability when donor funds decline or end. An evaluation of UNESCO's Community Multi-Media Centres initiative that included visits to 39 centers indicated that long-term benefits are being realized.[7] These include: gradual removal of barriers to social inclusion and poverty alleviation through access to knowledge of better health, resource management and agricultural practices. However, the evaluation notes, the tele-centers cannot survive solely on charges paid by individuals. This dilemma could mean that efforts to achieve financial sustainability will drive tele-center managers to target services for community members who can pay – to the detriment of those who are poor. Those managers need help in identifying suitable opportunities. Among the many *telecottages* in Hungary, there are more than 50 different kinds of services offered to the community. A major source is the contracts that their tele-center managers obtain from government agencies, thus becoming (for a fee) extensions for government services.

However, a key issue in the survival of tele-centers is the demand-driven factor, and that issue brings us to the consideration of reinventing universities.

Reinventing universities

Professor Calestous Juma of Harvard wrote in 2006 about the need to reinvent universities.[8] Although he was concerned especially with African universities he noted that universities worldwide were under pressure to expand their social mission. This theme has been struck in a variety of ways, ranging from the land grant university system in the United States with its associated "extension" programs (starting in the late 1800s and early 1900s), to recent international conferences targeting higher education and development – such as Higher Education and Participatory Development held in Vancouver in November 2006. The idea was on the agenda of the Asia-Pacific Programme of Educational Innovation for Development (APEID) at its 11th annual meeting in Bangkok in 2007. Its title was Reinventing Universities: Toward Participatory and Sustainable Development.[9] These forums touched on a wide range of issues related to the culture of universities but an underlying theme stressed the need for universities and other higher education institutions throughout the world to give higher priority to reaching out and serving local communities.

Higher education institutions and tele-centers

A recent (2010) analysis of learning centers in Nepal notes that the government and non-governemental organizations (NGOs) have established some 100 tele-centers throughout the country.[10] However, there is no evidence of systematic participation of Nepal's higher education institutions in the learning, information and training resources of these centers – beyond incidental technical assistance. Similar circumstances exist in other countries. The Vietnamese government's initiative to put Post, Telecom and Cultural Points (PTCPs) in communes around the country is a major part of developing the telecommunications infrastructure in rural and remote areas. The effort was intended to reach 35 per cent of Vietnam's 82 million people by 2010. These PTCPs are simple facilities or spaces within community buildings designed to provide basic post and telecom services, socio-cultural information, and specialized information to people in rural areas. Thus far, universities do not seem to have not been a significant part of the government's strategy to make these PTCPs a route for the Vietnamese people to become part of the information society.

There are clearly grounds for making a partnership between universities and community learning centers. For universities to become actively involved in ICT outreach such as supporting tele-centers, they must perceive an advantage in the arrangement. We think there are advantages and, thus, collaboration is mutually beneficial: universities and tele-centers have a logical affinity. In addition to the benefits that they can bring to tele-centers, the universities can benefit from an affiliation with tele-centers in at least three ways.

1. Tele-centers provide universities with a means for reaching beyond their "ivory tower" to extend their knowledge and learning resources to the surrounding communities and to other populations in the region. This includes translating, adapting, localizing and re-packaging information from external sources to fit the agronomic and cultural characteristics of those local communities. This function

is especially vital to the worldwide priorities identified in the MDGs. Ultimately, this makes universities and their faculties more relevant and better candidates for financial support from the public and private sectors and donors.

2. Tele-centers provide a laboratory for faculty and researchers to carry out ICT and extension-related research and development (R&D) projects especially involving issues ranging from HIV/AIDS to small business enterprises and poverty alleviation. Tele-centers as extensions of the classroom can also strengthen students' understanding of issues ranging from computer applications in community development to eGovernment and eCommerce. Graduates will be better prepared to enter a world that increasingly will be permeated with information and communication technologies. Initiatives such as those associated with eCommerce, eGovernment and eEducation will undoubtedly be a significant part of their lives and profes-sisons. Tele-centers provide a learning environment for students as tele-center volunteers to gain practical experience in helping people in the community – including grade-school and out-of-school youths – to apply information and communication technologies to challenges in their daily living. Students' assignments and semester-long or summer internships can be attached to courses in adult education, non-formal education, extension, communication and media, rural sociology, computer science and information technology and so on.[11] In some countries (e.g., Taiwan and Ghana), youths have a public service obligation for one or two years. While it is often associated with military service, attention can be drawn to adding community service in tele-centers as a means for discharging this obligation.

Both universities and tele-centers are involved in management of knowledge, learning and information resources. Specifically, universities can enhance tele-centers in at least four ways related to learning and information sharing.

- Contributing research: Research helps tele-centers become demand-driven. Research can identify communities' needs for information and related services. Tele-center staff generally have neither the skills, the time, the motivation nor the interest to carry out systematic research about the communities around them. Tele-centers also need research to evaluate continuously how well they are serving the needs of their communities.

- Providing local and relevant content: Much content on the Internet is not relevant to marginalized and disadvantaged populations. Institutions such as agricultural universities have access to science-based information that could be tailored to regional, provincial and local agronomic, social, linguistic and cultural characteristics. Also, universities are in a good position to design and administer distance learning for *non-formal* education.
- Training in information management: The staff in tele-centers need to be trained in how information can contribute to development, and how to help tele-center clients search for information as a reference librarian does in a library. Logically, universities have the capacity to teach and train, and these skills could be applied to these tele-center-related needs.
- Providing human resources: Tele-centers need volunteers who can help make them good places to visit, volunteers who can help people search and understand the basic rewards of a digital experience, and help those people navigate the various media in the tele-center. Female student volunteers can help draw village women into a more comfortable tele-center environment. Universities have students who could serve as tele-center interns.

The tele-center idea has caught on. There are thousands around the world. However, the idea of universities as partners of tele-centers has not caught on. As we have pointed out in the cases of Nepal and Vietnam, rarely will you find universities systematically linked up with tele-centers, despite the very clear benefit of a symbiotic relationship. Let me reach back a short time into India's digital history. In approximately 2003, M. S. Swaminathan, an outstanding leader associated with the Green Revolution in India, proposed a program called "Mission 2007: Every Village a Knowledge Centre." It would be an effort to provide rural communities with the power of ICTs. All 637,000 villages in India would be ICT-enabled – and use ICTs to address the major problems of rural India such as poverty, illiteracy, ill health and low farm productivity. The National Alliance, the executing body, had hoped to achieve its "Mission 2007" primarily through the creation of a network of Rural Knowledge Centres (tele-centers) across the country. When we proposed to an Alliance leader that the agricultural universities in India be explicitly included in the planning as partners for the knowledge centers, we received this terse response: "The universities have failed miserably in many respects. Most university faculty have no clue to life outside the campus nor have they any social concerns. Sorry for being very forthright or even blunt."

After 2007, "Mission 2007" was renamed The Rural Knowledge Movement and continued to aim for Every Village a Knowledge Centre. The Movement has more than 400 partners including international agencies, institutes, foundations, and local and national government bodies. It is notable that only two universities in one state appear on the web page list of partners.

There is some justification for the negative judgment about universities and their relevance to the real world. A two-year study commissioned by the World Bank and UNESCO a decade ago concluded that the contribution of higher education to social and economic development in developing countries has been "disappointing to date", including a failure to advance public interest. One of the major obstacles, according to the study, is that "the social and economic importance of higher education systems is insufficiently appreciated".[12]

Beyond the rhetoric

We return to the judgment that universities, especially in developing nations, are irrelevant. We need to emphasize cases in which universities *are* relevant. Here are three, the first two from India, and the third from Africa.

The first is a 1969 initiative in which the Ford Foundation helped build the capacity of the G. B. Pant University of Agriculture and Technology (then in Uttar Pradesh) to produce radio programs for Indian farm families. Radio was the major "ICT" of that time. In producing complete radio programs and delivering them on tape to the All India Radio (AIR) station in Rampur, the university became the first non-Indian government body to provide radio programs on AIR for public consumption. Was it sustainable? Now, more than 40 years later, according to our correspondent in Pantnagar, the University produces 20–25 programs each month, which are broadcast regularly by two stations of AIR. When there is any significant research information or any new agricultural variety released, it is sent to the entire network of AIR (comprising 165 stations), especially in India's northern states. The university at one point was selected by FAO to be a Centre for Advanced Study in Agricultural Communication.

The second case comes from Tamil Nadu, where support from Canada's International Development Research Centre and UNESCO increased the capacity of the Tamil Nadu University of Veterinary and Animal Sciences (TANUVAS) to incubate and support three Village Information Centres (VIC) and six additional information centers especially serving Self Help Women's Groups. All were community-driven and some were fully supported by their surrounding communities. The university continues to play a partnership role through some training and advisory initiatives and by supplying content relevant to the various micro-enterprises in rural Tamil Nadu. For example, it has produced video CDs on subjects ranging from remedial measures for infertility in cattle to "Cream, Butter and Cheese" – all designed for rural populations and available through the tele-centers.[13]

In Africa, according to Professor Juma, the University of Zambia was the midwife of Zamnet, the country's largest Internet provider. It is a "shining example", says Juma, of business "incubation" by a university.

An ICT for development (ICT4D)project in northern Vietnam adopted a model unique among tele-center projects around the world. Funded by the APEC Education Foundation (AEF) in 2006, its most prominent feature is a cluster of community-based and community-operated tele-centers incubated by a university and continuously supported by the university with content, research and training resources. The ICT Project, as it was named, took place in the Northern Mountainous Area (NMA). The NMA consists of 16 provinces. It covers an area of 110,000 sq km (34 per cent of the area of the whole country). It has a population of approximately 16 million people (20 per cent of the population of Vietnam). Thirty five of Vietnam's 54 ethnic minority groups reside in the area. Due to the difficulties in the region, the NMA is considered the poorest and most disadvantaged region of Vietnam.

The pilot phase was designed to demonstrate the viability of Vietnamese universities as incubators and institutional supporters of community information centers (tele-centers), to create a communication network for exchange of information, and to provide IT education for communities, particularly for disadvantaged youths in the northern mountainous region of Vietnam. Also, the project attempted to help the communities understand how ICTs could benefit them.

Steps toward university–tele-center partnerships

The Action Plan of the World Summit on the Information Society (WSIS) called for creating a critical mass of skilled ICT professionals and experts dedicated to development-related issues and institutions. Yet a United Nations press release noted that recent research has indicated that university and other higher-learning institutions in the Asia-Pacific region responsible for training future leaders and policy makers lack adequate academic content on ICT for development in their curricula. Programs and courses that are best suited to provide training and impart knowledge about the use of ICT for socio-economic development either do not cover information and communication for development or address it in a manner that does not sufficiently identify the potential of ICT.[14]

We have noted elsewhere the importance of building human resources to take advantage of information and communication technologies and some promising steps that are being taken.[15] For example, the Asia and Pacific Training Centre for Information and Communication Technologies for Development (APCICT) recognized that in an increasingly digital and inter-connected global environment, policy makers and many other government officials are indispensable in creating and sustaining an environment that promotes the use of ICTs for development. The organization recognizes that policy makers set priorities, procure and allocate funds, facilitate alliances and partnerships, and enact policies that promote the ICT industry sector and applications of ICTD. So APCICT created an impressive "academy" to train government leaders in the region that involves eight comprehensive modules on a variety of relevant ICTD topics, international and national workshops, and a distance learning scheme for delivering the modules. The Centre especially focuses on building the capacity in its member countries to continuously sustain the building of ICTD-skilled human resources.

The Centre has also looked to the more distant future by engaging universities in the ICTD movement. They organized a November 2010 Consultative Workshop in Inchon, Korea, on "Turning Today's Youth into Tomorrow's ICTD Leaders." Seventy five policy makers and experts associated with higher education institutions from 29 countries in the region attended. A major focus was on university curricula and especially on the need for higher education's academic programs to contribute to using ICTs for socio-economic development in the Asia-Pacific region. The event was called "an important step" in APCICT's "Strengthening ICTD Education in Institutions of Higher Learning" project.[16] The project aims to raise the awareness of policy makers, governments and a network of partners in the region about the importance of ICT for development education in tertiary schools. The organization will build an ICTD university learning module that can be used at institutions of higher education to either provide core training or supplement existing course materials. The module was scheduled for launch in October 2011.

We see how to operationalize the idea of ICTDs in a university curriculum by looking at some of the activities in the Vietnam project where an early step was made to build ICT activities into the curriculum. Some students at the Thai Nguyen University of Agricultre and Forestry

(TUAF) were sent to tele-centers in the communes at graduation time to work for six months as commune extension workers. Also, undergraduate and graduate students in the Department of Extension and Rural Development spent two to six months working on research related to communication and development; some worked on surveys or research conducted by TUAF staff. Incentives for students to become involved in these ways included merits in their graduation examination mark for field activities. Some research projects provided a little allowance supporting the students in the field.

Faculty participation in ICTD

It is important that university faculty members support innovations such as ICTD. Unfortunately, higher education institutions often have faculty members who survive on lecture notes that are out of date, and innovative young members are inadequately recognized. Universities need to support and reward faculty who step outside the historical expectations of their institutions. TUAF took steps to involve faculty and staff in the tele-center project. The university set up a team, which consisted of a number of scientists in various fields, to be responsible for producing content for tele-center clients. Some of this was deliverable online; significant, however, were materials produced for consumption offline in the tele-center facilities, via CDs and DVDs. TUAF also designated two staff members to provide information technology technical support and training for tele-center operators. From the support of the university-led ICT4D project, each tele-center consisted of an information resource of more than 54,000 information products that included: documentary and technical films, text book material, technical articles, technical guidelines and so on. But this content mainly focused on agriculture, animal husbandry, aquaculture and rural development. TUAF staff perceived the necessity of developing more content to cover other sectors such as health, education, market information and natural resource management. An important step for both universities and tele-centers is to build library resources into the partnership with a reference librarian as a consulting partner.

Conclusion

The APCICT project described above recognized the importance of having policy support for introducing and enhancing the ICTD curriculum. In promoting a university–tele-center relationship, policy needs to extend beyond the curriculum. There needs to be policy support for university outreach, similar to the legal framework for the university support of extension in some American states (such as New York). In Vietnam, TUAF had a mandate to offer formal and non-formal education for communities in the Northern Mountainous Area of the country. Thus, tele-centers are perceived by the university administration as community-based learning centers for local people using modern information technology. It was consistent with this mandate to use the university's human resources to provide technical support and educational resources via the tele-center system.

Universities need to provide appropriate on-campus ICT laboratories for training students, for related ICTD research, and for the development of ICT content suitable for the needs of the rural communities. However, given the common constraint of maintaining and upgrading such resources, universities and tele-centers need to identify mechanisms for continuous funding of facilities – such as fostering ICTD related micro-enterprises. TUAF had an initial grant from the Education Foundation of Asian Pacific Economic Cooperation (APEC) and an IT equipment company to establish its laboratories and equip six tele-centers. However, the lack of support from the Vietnamese Ministry of Communication and Telecom appeared instrumental in the decline in the condition of the ICT facilities, thereby contributing to a general degrading of services.

It is important to note that an active, visible and successful university ICTD program can have two additional outcomes. One is the simultaneous building of the university's own internal ICT infrastructure which will contribute to the quality and efficiency of the university's academic and administrative functions. And second is the reshaping of its relationship to the outside world – being perceived as a more active agent of change. This addresses the observation of the Task Force on Higher Education and Society that "Unlike primary and secondary education, there is little in the way of a shared vision about the nature and the magnitude of the potential of contribution of higher education to development." Surely it is an imperative for the future.

Notes

1. "Twitter tweets are 40% 'babble' ", *BBC News*, 17 August 2009. Retrieved 26 July 2011, from http://news.bbc.co.uk/2/hi/8204842.stm
2. Colle, R. (2006). Erskine Childers: A devcom pioneer. In Gumucio-Dagron, A. & Tufte, T. (Eds.), *Communication for Social Change, Anthology: Historical and Contemporary Readings* (pp. 495–506). South Orange, NJ, Communication for Social Change Consortium.

3. Fraser, C. & Restrepo-Estarda, S. (1998). *Communicating for Development*, London, I. B. Tauris, p. 39.
4. The background on the G8 initiative can be found at: The Commonwealth Government and Business Guide to Information and Communication Technology, *Digital Opportunity Task Force*. Retrieved 26 July 2011, from http://www.cto-ict.org/index.php?dir=07&sd=40&org=int&oid=1007
5. *The Grameen rural phone system once numbered 270,000 "telephone women" in 50,000 Bangladeshi villages but one report notes that they no longer exist. See: GP's Palli-phone women are gone*, Bangladesh's Telco News, 5 December 2010. Retrieved 26 July 2011, from http://www.facebook.com/note.php?note_id=171234876244754
6. Kenny, C. (2006) *Overselling the Web, Development and the Internet*. Boulder: Lynn Rienner.
7. Creech, H., Berthe, O., Assubuji, A.P., Mansingh, I. & Andjelkovic, M. (2006). *What We've Learned, Key Findings from an Independent Evaluation of UNESCO's Community Multimedia Centre Initiative*. Paris: UNESCO.
8. Juma, C. (2006). Reinventing African universities. In *Falmer* (the University of Sussex Magazine), 44, 80–10 (Summer).
9. For a summary of the proceedings and discussions at that conference see *Reinventing Higher Education, Toward Participatory and Sustainable Development* (2008). Bangkok: UNESCO.
10. Tulachan, A. (2010). Telecenters as local institutions: Present landscape in Nepal. *The Journal of Development Communication*, 21(1), 62–68.
11. Colle, R. (2005). Building ICT4D capacity in and by African universities. In *International Journal of Education and Development using Information and Communication Technology (IJEDICT)*, 1(1), 101–107. Retrieved 26 July 2011, from http://ijedict.dec.uwi.edu/viewissue.php?id=3
12. Task Force on Higher Education and Society (2000), *Peril and Promise: Higher Education in Developing Countries*, The World Bank, Washington, DC.
13. Colle, R. & Roman, R. (2003) "University-Based Telecenters". In Badshah, A., Khan, S. & Garrido, M. (Eds), *Connected for Development: Information Kiosks and Sustainability, Series 4* (pp. 89–96). New York: UN ICT Task Force.
14. APCICT Academy Background. Retrieved 26 July 2011, from http://www.unapcict.org/academy/background
15. Colle, R. (2011). "Building Leadership in the eWorld", *The Journal of Development Communication*, 22(1), 1–8.
16. APCICT Future Leaders ICT Programme. Retrieved 26 July 2011, from http://www.unapcict.org/aboutus/programmes/advisory/future-ict-leaders-programme

Part III

Communicating the Cost of Social Change

7
Communicating the True Ecological Cost of Development: Addressing Development and Environment in Orissa, India

Maitreyee Mishra[1]

> You must teach your children that the ground beneath their feet is the ashes of our grandfathers. So that they will respect the land, tell your children that the Earth is rich with the lives of our kin. Teach your children what we have taught our children – that the Earth is our mother. Whatever befalls the earth befalls the sons of the earth. If men spit upon the ground, they spit upon themselves.
>
> Chief Seattle of the Duwamish Tribe
> to the new "Americans"[2]

Introduction

Orissa,[3] located in eastern India, is a land of contrasts. It has a history dating back to over 2,000 years, with stories of trade with South East Asia during ancient times. Endowed with rich natural resources, a long coastline and abundant forests, it is also culturally diverse. Apart from mainstream Oriya culture and traditions, interior and western Orissa are inhabited by hundreds of tribal peoples, each with their own culture, gods, traditions and practices. In a land where all beliefs need to be respected and uniqueness celebrated, unfortunately, the *adivasis* or the tribal peoples have always been the "other." As it is with all cultures, these peoples form the periphery, with alien customs that must be integrated with mainstream culture and society. The dominant political philosophies in the post Second World War period upheld the importance of westernization, industrialization and modernization, and hence the need to "develop" all peoples. Indigenous peoples, in all parts of the world "found themselves increasingly incorporated within the world economy and subjected to national development goals and programs" (Wong, 2008). In Orissa, as in other parts of India, the state and country's needs remained superior and hence lands were acquired from many *adivasi* communities, displacing them.

Over the last 50 years or so, modernization has instilled a new culture in Orissa, as elsewhere. The pressures of development have been immense and so has the need to intensify economic growth, abandon traditional beliefs and customs and, in turn, ignore the needs of the environment. Neoliberal reforms in the 1990s have further intensified the need for economic growth. Orissa, though rich in natural resources and culture, is one of India's poorest states, with a per capita Net State Domestic Product in 2008/2009 of Rs. 29,464 (Reserve Bank of India, 2010), which is 24 per cent below the all-India average. Observers have claimed that the state's poverty is a result of several factors, some of which could be due to the political establishment and lack of political will. Many strongly argue that its poverty is a result of the changes and variability in its climate. The last 50 years, in particular, have witnessed a rapidly changing climate and ecology. The dominant view is that poverty is a result of under-utilization of the state's vast natural resources, while some serious thinkers believe that this would in turn lead to ecological destruction and irreparable damage to the very people who inhabit those lands containing the vast natural resources.

This chapter looks at the dichotomy of the complex development–environment debate in Orissa, with its various contrasts, as will be discussed later in the chapter, namely: rich resources, poor state, ecological crisis, government focus on industrialization and extraction of mineral resources, and the environmental and human costs of all industrial and development activities. Through three case studies, this chapter looks at indigenous peoples in Orissa who have been at the receiving end of all development and environmental conflicts, displaced from their lands, their livelihoods snatched and their gods and beliefs belittled. In the process it addresses the recent grassroots movements and uprisings against forced displacement and destruction of their homes. This chapter also looks at the kinds of communication, dialog and networking evolving from these tribal peoples' movements; and studies the role of the mass media in these debates.

Orissa's erratic climate and environmental change

The last few decades have seen huge changes in Orissa's climate. Observers and people living in the state have noticed that the weather is no longer predictable and that the seasons have changed such that spring and winter have almost disappeared. This means that the monsoons are erratic and the summers are long and excruciating. Even coastal Orissa, which used to have moderate weather during summer, now witnesses extremes in temperatures, which can rise up to over 40 degrees Celsius during April and May.

Rising temperatures and erosion of the seasons are not the only climatic changes Orissa experiences. The state has been witnessing alternating droughts and floods, cyclones and heat waves together with extensive coastal erosion, which is very evident in coastal areas such as Puri, Jagatsinghpur and Kendrapada. Use of pesticides and fertilizers during the Green Revolution vested on increasing agricultural output has been a contributing factor to agricultural decline, with the loss of traditional farming systems. In 1998, over 1,500 people were killed due to a heat wave (Agarwal, Sharma & Narain, 2001). The frequency of cyclones has also increased. 1999 witnessed the highly destructive "super cyclone", one of the worst in hundreds of years, which lasted for almost three days. Shiva (2002) notes that the 1999 cyclone, which caused the deaths of over 20,000 people, destroyed 1.83 million houses and uprooted 80 per cent of coconut trees, was a "man-made ecological crisis unleashed by the combined impact of climate change, industrialization and deforestation". Two years after the cyclone, in 2001, Orissa was faced with a severe drought, followed by floods during monsoons that affected 7 to 11 million people (Shiva, 2002; Agarwal et al., 2001). The cycle of drought and flood has been replayed since then, with floods in 2008 affecting over 600,000 people. 2010 was also a year of droughts due to delayed monsoons (rainfall deficit was almost 35 per cent around July), resulting in paddy crop loss of 50 per cent (Patnaik, 2010; NDTV, 2010). This was followed by flash floods in August 2010, affecting some 85,000 people, mostly in Southern Orissa (Outlook India, 2010).

These climatic changes have contributed to loss of paddy crops (rice being the main agricultural crop in the state). The district of Jagatsinghpur in coastal Orissa was one of the worst affected due to the super cyclone. Today, it has the lowest forest cover in the state (Forest Survey of India, 2009; Forest Survey of India, 2011). Mangroves, which can help reduce the impact of cyclones, have been systematically destroyed for aquaculture, further aggravating the situation (Shiva, 2002; Mahapatra, 2006).

The government of Orissa's Climate Change Action Plan (2010) states that the government wants "to ensure that climate change does not undermine the economic development which is now underway". It also stresses the importance of economic growth and "increased emphasis on power, mining, energy intensive industries and infrastructure" (Government of Orissa, 2010). Ironically, the same report recognizes that "mining projects are threatening forests, livelihood of people dependent on forest based economy" with a high carbon footprint (Government of Orissa, 2010). Despite the impending ecological crisis, the Orissa government has been bent on economic growth based on extraction of mineral resources.

Industrialization and the state

Since the 1950s the various governments in the state have focused on industrial and agricultural growth based on intensive inputs. The Green Revolution furthered the notion of industrialization of agriculture, eventually leading to the abandonment of traditional practices in order to increase agricultural output. Industrial growth remained in state control until the 1990s, when liberalization and economic reforms stepped in and put emphasis on privatization. Post-liberalization, the shift has been on extraction of Orissa's rich natural resources by opening up the state's resources to private players, both national and international. The paradigm of quick industrialization is largely supported by most political parties, the middle classes and the mainstream media.

For the state of Orissa the pressures of competing with other states in a globalized Indian economy have been immense and the government has largely looked at the inflow of investment. Over the last several years, the government has signed 49 Memoranda of Understandings (MoUs) with national and international industrial houses to set up industrial plants across the state (Government of Orissa, 2008; Amnesty USA, 2007). These industrial houses include the United Kingdom's Vedanta Resources Inc., and Rio Tinto, Canada's Alcan, South Korea's Pohang Steel and Iron Company (Posco), Arcelor-Mittal, Jindal, Tata and Hindalco, making it approximately 2,000 billion rupees ($ 37 billion) worth of investment proposed, of which 166 billion rupees ($ 3 billion) have been invested so far (Government of Orissa, 2008). Some larger projects that are underway are Vedanta Resources bauxite plant at Lanjigarh in Kalahandi district, Posco's steel plant at Paradeep in coastal

Orissa, Tata's steel plant at Kalinganagar and Arcelor-Mittal's steel plant at Keonjhar.

The reason for such investment and investment promises lie in Orissa's rich bauxite, chromite and iron reserves. 60 per cent of India's bauxite reserves are located in Orissa; the state also has large reserves of iron ore, chromite and coal (Government of Orissa, 2005). Orissa has a forest cover of 31.38 per cent (Forest Survey of India, 2009); most of it is well spread out except in the coastal belt. Most of the mines are located in the forest areas, regions that are inhabited by numerous tribes. The areas mentioned above: Lanjigarh, Kalinganagar and Keonjhar are in the forest belt, with a concentration of tribes who are at risk of being displaced from their ancestral lands.

The state and Orissa's tribal communities: past and present

The government's focus on rapid industrialization has been based on extraction of mineral resources. The state-controlled Industrial Infrastructure Development Corporation of Orissa (IDCO), established in 1981, has the responsibility of acquiring lands to develop industrial infrastructure so that foreign and national industrial houses would be interested in investment in Orissa. This has been a further developed role post-liberalization, where the corporation acquires land and sells it to private players. It has been doing so and, as discussed later in the Kalinganagar case study, this has posed certain problems.

Most mineral resources are located in interior areas of Orissa, which are ecologically pristine, inhabited by several different tribes, each with customs and cultures that are intrinsically unique to that region. These tribes have gods and beliefs which are different to mainstream religious beliefs and traditions in Orissa. The Dongria Kondh, for example, that live in the Niyamgiri hills believe that their god, Niyam Raja created the Niyamgiri hills for the tribe to inhabit.

Much like the beliefs of the Dongria Kondh, other tribes of Orissa have an innate connectedness with their environment, worshipping the elements of nature and cohabiting with all creatures, believing that the hills, forests and rivers are their gods, siblings and parents. The very essence of their existence is in connection with these elements of nature. As Wong (2008) observes, "for indigenous peoples, land has a sacred quality, is revered and respected and is inalienable".

The opening up of Orissa's natural resources has both ecological and social implications. The various industrial plants spread across the state have polluted rivers, streams and wells that are the lifeblood of tribal peoples who have depended on these sources of water for generations. The setting up of industrial plants in these regions has also led to displacement of tribal peoples. Displacement itself comes with both visible and invisible costs. The visible costs can be addressed perhaps with monetary factors. The invisible costs, however, which include emotional, cultural, social and spiritual dislocation, are complex, incalculable and cannot be repaired with any material replacement.

Such displacement often creates a state of vulnerability for these peoples, as their sense of self and their identities are uprooted, and they are reduced to the status of outsiders in the land to which they belonged and that they knew and understood. They are further dislocated through displacement from "resource-rich" to "marginal lands" (Wong, 2008).

Stories of displacement are not new in Orissa. Hirakud Dam, constructed in 1956 in Orissa's Sambalpur district submerged 249 villages in the district (Baboo, 1997). In 1945, when the dam's site was decided, the people of Sambalpur launched a campaign against the dam to protest against the destruction of these villages and cultivable lands (Baboo, 1992). Nanda and Tripathy (1987) note that people's protests were dealt with by force, and by persuasion through compensation, on the basis of land for land and house for house by the then-ruling Congress government (cited in Baboo, 1997). Baboo (1997) observes that the government had promised rehabilitation in the form of colonies but people were not interested in moving to the camps; displaced people had the option of either moving to "old established villages" or remaining in partly submerged villages. He notes that resettled tribal people, who had been given very little compensation from the government, suffered from not being able to adjust to their new surroundings and could not get employment. Those who had shifted to already existing villages continued to be viewed as outsiders. Other incidents of displacement include tens of thousands of people displaced as a result of the steel plant constructed at Rourkela, after India's independence.

Since the 1990s, with the increase of mining and industrial activities across these tribal belts, many communities have been displaced. With the intensive liberalization of Orissa's economy and the extensive inflows of investment, there has been a groundswell of resistance against such displacement by several *adivasi* communities, grassroots organizations and an interwoven network of national and international support organizations and human rights groups.

The Forest Rights Act (FRA), which was passed by the Indian parliament in December 2006 and came into force in December 2007, makes it mandatory to receive the consent of the village Gram Sabha[4] for any land acquisition (Hindustan Times, 2010; Government of India, 2006). The Forest Rights Act brought some relief to tribes and other forest dwellers across the country whose rights to the land which they have inhabited and depended on for generations were reinstated.

Many communities across Orissa, particularly tribal peoples, are voicing their dissatisfaction at the destruction of the environment and forced displacement from their ancestral lands as a result of mining and development activities. The following three case studies discuss people's movements against displacement and ecological change – Tata's plant at Kalinganagar; Posco's proposed steel plant at Paradip and United Kingdom-based Vedanta Resources' bauxite plant at Lanjigarh/Niyamgiri.

Case study 1: Kalinganagar

The Kalinganagar Industrial complex has its beginnings in the 1990s' economic reforms. Strategically advantageous, Kalinganagar is located near iron and chromite mines and has the highest concentration of manganese and iron (Pandey, 2008). Kalinganagar industrial complex spans some 12,000 acres of land and covers 83 revenue villages (Pandey, 2008).

The Industrial Infrastructure Development Corporation of Orissa (IDCO) was given the task of acquiring land and developing infrastructure for the proposed industrial complex at Kalinganagar in order to attract investors (Pradhan, 2006). IDCO started acquiring land in the years 1992–1994 and had purchased land from people with proper title deeds (patta) at the rate of Rs.15,000 to Rs. 30,000 per acre in the initial acquisition phase (Pradhan, 2006). As the tribal people who had been cultivating most of those lands for generations had no title deeds, most of the land acquired was dubbed as government land. Those with deeds were compensated; a huge proportion of people were left uncompensated (Pradhan, 2006).

Some 12 large and small industrial houses were allotted land in Kalinganagar. This proposed industrial area has an average concentration of 30 per cent of tribal peoples (Pandey, 2008). Pradhan (2006) observes that IDCO sold these acquired lands to companies at higher prices: land was sold at Rs. 350,000 per acre to Tata in 2004–2005; the company itself was allotted 1,960 acres of land (Pandey, 2008).

The inadequate compensation and problems in addressing rehabilitation issues caused agitation among the people. They continued to occupy and cultivate their land as they had not been compensated and had no alternative livelihoods. Pacification efforts by the government included some additional compensation (Rs. 25,000 per acre) but when the people discovered that their lands had been sold to industries at higher prices, they were further angered (Pandey, 2008).

In 2004, the organization, Visthapan Virodhi Jan Manch (VVJM) was formed to represent the people affected by the Kalinganagar Industrial Complex. In October 2004 they wrote an open letter to the Chief Minister of Orissa informing him of their concerns and hardships and laid down demands for displaced and to-be-displaced people; these included:

- halting construction in agricultural land;
- lease deeds to be given to those settled before 1980; and
- one job per displaced family (Pradhan, 2006).

In October 2005, Tata arrived at the location to construct a boundary wall, resulting in strong opposition from the people, who demanded that they be rehabilitated before being displaced. Tata postponed the construction till 2nd January 2006, when they returned to complete the boundary wall, with the support of the administration and deployment of paramilitary forces. There was opposition from the tribal people and this resulted in a clash between the forces and the tribal people, resulting in the deaths of 12 tribal people (including one child) and one policeman (Down to Earth, 2008; Pandey, 2008; Times of India, 2006).

The killings at Kalinganagar drew national and international attention to the problems in the region and there was condemnation of the event. The protesting *adivasis* are still awaiting justice for the killings (Amnesty International, 2012c). The killings further provoked the tribal people and they blocked the Daitari-Paradip express highway which is "considered an economic lifeline" linking the mineral-rich Jajpur and Keonjhar districts with the port at Paradip (Down to Earth, 2008). Pandey (2008) notes that the killings have made the tribal people even more determined that the government meets their demands, which include that there be no further displacement and "five acres of land in lieu of land acquired" be given to them, together with compensation for the families of the dead and injured.[5] The agitations have continued at a lower intensity, and the government has responded by isolating those who were protesting.

In April 2010, the Congress and the Bharatiya Janata Party (BJP) went on a fact-finding mission, accusing the state government of "thrusting upon industrialization on tribals" (The Hindu, 2010a). In May 2010, another round of clashes, which involved 1,000 police officials against 300 *adivasi* protestors, resulted in the killing of a 50-year-old *adivasi* leader, Laxman Jamuda, while ten protestors were injured (Amnesty International, 2010b). Even though construction work is ongoing, in April 2012, it was announced that Tata Steel would commission the first phase of its Kalinganagar operations in March 2013 (The Economic Times, 2012). However, this has been deferred and operations are expected to start by 2014–2015 (The Hindu, 2012).

Protests have continued in Kalinganagar, mostly demanding better compensation packages for displaced families. In September and November 2012 there were further demonstrations by the displaced people.

Case 2: Posco

Pohang Steel Company (Posco), the South Korean steel company, signed an MoU with the government of Orissa in June 2005 for a steel mill project at Paradip, in Jagatsinghpur district, with the construction of the plant to begin in 2011 (Mukherjee, 2010). The site chosen for the project was to displace more than 2,000 people. This triggered fear in the hearts of the villagers who lived at the proposed site over losing their livelihoods and subsequent environmental destruction. For the last seven years, the people of three *gram panchayats*[6] in Jagatsinghpur district have been protesting against what they deem an "illegal" occupation of their lands. Later in 2005, the people's organization, Posco Pratirodh Sangram Samiti (PPSS) was formed against the proposed Posco project. In 2006, the Forest Rights Act was passed. The PPSS held that under the Forest Rights Act (FRA), the lands of these villagers cannot be acquired without their consent as they fall under the "other forest dwellers" category under the FRA. The PPSS called for negotiations, and declared a blockade in three *gram panchayats* (Campaign for Survival and Dignity, 2010). In November 2007, the PPSS dharna (protest) was attacked by the police (Campaign for Survival and Dignity, 2010). In March 2008, Dhinkia[7] Gram Sabha passed a resolution for recognition of its rights under the Forest Right Acts (Ghosh, 2010).

In December 2009, the Forest Advisory Committee gave the approval for Posco's steel plant (Hindustan Times, 2010). Of the 4,000 acres of land required by the company, 2,900 acres is forested area (Mukherjee, 2010). In May 2010, the Orissa government refused to shift the site of Posco's plant and asked the villagers to reconsider the rehabilitation package (The Hindu, 2010b).

The protests continued and there were more attacks on protestors in May 2010, where several protesters were injured (Campaign for Survival and Dignity, 2010). In August 2010, the acquisition work was halted as a result of a report by a central government committee (NC Saxena Committee) to look at the implementation of the Forest Rights Act (Hindustan Times, 2010). In November 2010, this committee "confirmed" that the Forest Rights Act had been violated by the Orissa government through the acquisition of land for Posco's project. However, in May 2011, following the state government's claims that there are no forest dwellers in the area, and hence no settling of rights under the FRA, the project was given environmental clearance by the Ministry of Environment and Forests (MoEF). Since then, there have been continued protests by villagers and the PPSS in various forms, including human barricades (Bera, 2011b). Protesters, however, have not been dealt with kindly: stories of fabricated cases filed against those protesting against the Posco project are common, and many are faced with multiple registered cases; some have also been jailed while others have been labeled Maoists (Maneckshа, 2012). November 2012 saw tensions between the pro-Posco and anti-Posco groups. On 6 December 2012, the Indian government refused to extend the in-principle approval given to Posco in 2006 for the setting up of a multi-product Special Economic Zone (SEZ).

Case 3: Niyamgiri and Vedanta

Vedanta Resources, a UK-based mining company signed an MoU with the government of Orissa in 2003 to construct an alumina refinery and coal thermal plant at Lanjigarh in Kalahandi district. The Ministry of Environment and Forests (MoEF) gave environmental clearance to the company to mine on the basis that it would not use forest land. The company built a refinery at Lanjigarh in 2006. The refinery caused the displacement of over 100 families (Survival International, 2010a). Lanjigarh is located at the foot of the Niyamgiri hills, inhabited by the Dongria Kondh tribe. Vedanta wanted to set up an open cast mine in the Niyamgiri hills by blasting the top of the mountain to extract several million tons of bauxite. Its activities were supported by the government of Orissa, who saw this as an opportunity for economic growth.

For the Dongria Kondh tribe, the Niyamgiri hills are sacred. These *adivasis* collect bananas, roots, vegetables, palm juice and so on from this mountain, depending on forest resources for livelihood. The *adivasis* rose to defend Niyamgiri from Vedanta by blocking roads to their sacred hills. Members of the tribe appealed against Vedanta Resources to prevent the mining company from destroying their sacred mountain and forests. There were "symbolic gatherings" of the tribe on the top of the hill, ritualistic singing and other traditional methods of protest, some of which were covered by national and international media (Mishra, 2011). Their voices were carried across by grassroots organizations such as Niyamgiri Suraskha Samiti and international organizations such as Survival International, Action Aid and Amnesty International. The work of these organizations and the information that flowed about the community's struggle helped to draw the attention worldwide, with local, national and international media being drawn into the discourse.

Amnesty International blamed Vedanta for violating "human rights to water and health" as a result of poor waste management and pollution from Vedanta's refinery (Amnesty International, 2010). The sources of water were also mixed with bauxite dust (Survival International, 2010a).

In August 2010, the tribe won a victory against Vedanta when India's government declared that Vedanta would not be allowed to mine in the Niyamgiri hills (Survival International, 2010b; Rehman, 2010). In October 2010, the Indian government blocked Vedanta from expanding its alumina refinery at Lanjigarh, at the foot of the Niyamgiri hills for violating the Environmental Impact Assessment (EIA) notification (Times of India, 2010).

In August 2012, Amnesty International accused Vedanta of not making progress with respect to human rights and of not consulting with or disclosing important information regarding the impact of pollution caused by the activities of the company, with the concerned communities (Amnesty International, 2012b). It also stated that "those opposed to the company's operations have faced fabricated charges", also referring to the ongoing inquiry by India's National Human Rights Commission, which reveals that the state police have promoted the company's interests by "framing of false charges and in the suppression of dissent" (Amnesty International, 2012b).

Having been denied access to the Niyamgiri hills, the company had been reliant on bauxite stock from the states of Gujarat, Chhattisgarh, Jharkhand and Andhra Pradesh, though this proved to be difficult with regulatory, logistic and procedural issues getting in the way (Satapathy, 2012). In September 2012, the Vedanta group announced that they would shut down the alumina refinery at Lanjigarh in December 2012 as a result of bauxite stock depletion (Press Trust of India, 2012); the refinery was finally shut down on 5 December 2012.

As of December 2012, although protesters are happy with these developments, there is still a great deal of uncertainty regarding the permanent closure of the refinery. In March 2011, the Orissa Mining Corporation, which had signed a deal with Vedanta, moved the Supreme Court against the Ministry of Environment and Forests' (MoEF) cancellation of permission to mine the Niyamgiri hills; this case is pending and the hearing is scheduled for January 2013. There is mounting pressure on those employed in the refinery to oppose the shutdown, with many fearing the loss of their jobs. Activists such as Prafulla Samantara feel that Vedanta is putting pressure on the government and garnering public opinion in its favor in view of the Supreme Court hearing.

Network of people's voices

The three case studies, although different, are each illustrative of people's struggles against large corporations. These are struggles against the stripping away of tribal identity, ways of living and livelihood. In Orissa, where areas inhabited by tribal people are rich in natural resources, development has meant the encroachment of lands held for generations leading to physical, emotional and spiritual displacement and dislocation of communities.

Unlike the 1950s and 1960s, when minority community interests were not publicized and were ignored to make way for larger national development goals, the last two decades have seen a shift in both awareness and empowerment of marginalized communities. People are now resisting change and standing up against forced displacement, and destruction of the ancestral lands that they have inherited and inhabited for generations, in both physical and spiritual forms. These grassroots movements against environmental change are showing us that the earth comes before the material benefits that development offers.

While communities have been able to express themselves, there is also a greater understanding of the needs and requirements of these communities, which are at the receiving end of environmental change. The central government's receptivity to tribal communities has definitely seen an improvement; in the case of India the Forest Rights Act was a step forward. The mainstream perceptions of indigenous people have changed from destroyers to protectors of the environment.

The struggles of the Dongria Kondh are based on their intrinsic relationship to the Earth. This relationship, based on sustainability, explains their rush to action at the possible destruction of their sacred hill. The tribal movement itself led to blockades of their sacred hills, the tribe's symbolic gatherings on the top of the hill, ritualistic singing: some of these traditions were mentioned in newspapers, both national and international. The Dongria Kondh were able, through the sustained work of grassroots organizations such as the Friends Association for Rural Reconstruction (FARR), to understand their rights and the potential damages of Vedanta's actions and to voice their grievances at the initial stages. This probably laid the foundation for the networks that formed at the local, national and international levels. At the local level, the Niyamgiri Suraksha Samiti (Niyamgiri Protection Committee) was formed, and other non-governmental organizations (NGOs) and organizations joined in; these groups that dealt with human rights and environmental issues were instrumental in filing the case against Vedanta in India's Supreme Court (Xaxa, 2012). Later, several other groups, activist organizations, NGOs (both national and international), media organizations and individuals stepped in, helping extend the voices of the Dongria Kondh and taking them to another level. Survival International, Amnesty International, and other organizations stepped in to support the tribe; celebrities also showed support. The mainstream media, too, found greater cause to cover the culture, spirituality and environmental unity of this tribal group. International newspapers and media also moved in to cover the story by traveling to the Niyamgiri hills. Niyamgiri's case drew a lot of attention internationally with the building of grassroots, local, national and international networks; advocacy by activist organizations, and perhaps also leading to an appreciation and greater understanding of the environment and of the customs and traditions of the Dongria Kondh. Thus, through this coverage, through the public outcry and advocacy provided by organizations such as the ones mentioned above, an international public opinion was shaped. Information was available on the Dongria Kondh, and people all over supported the tribe. At the state level, as the Orissa government's sympathies lay with Vedanta, receiving environmental justice was a challenging task. Vedanta's Orissa projects and its marketing of probable benefits to rural communities in Orissa are well communicated (or perhaps miscommunicated) through the company's huge hoardings that were spread across Orissa's capital Bhubaneswar and other cities for several years, saying "mining happiness". Many groups of people in Orissa's capital, Bhubaneswar saw the agitation of the tribal people at Niyamgiri to be baseless and undermining of Orissa's forthcoming economic growth promised by Vedanta. These arguments are manifested in most of the media, which, particularly in the case of Kalinganagar, have not been sympathetic to the people's struggles.

Communication has the potential to shape our understanding and perceptions of the environment and thus define how we behave toward it (Milstein, 2009). The mass media in general, though somewhat receptive to the needs to these communities and of the environment, have not been very representative of these counter-development arguments. This is more so among the mainstream media in Orissa's print and television, where there is a heavy reliance on government sources of information due to their political lineages and hence little alternative information is available on environmental damage (Mishra, 2008). However, there has been some coverage given to the Niyamgiri-Vedanta issue on the Internet by alternative media organizations, and thus there are emerging models of communication in horizontal patterns. Following the cancellation of Vedanta's environmental clearance, the mainstream media too seemed to have been drawn into the discourse.

There are some inherent similarities between the three cases mentioned above. Communities that have been held together by a common history, tradition and culture have been pitted against each other, owing to the complexities of the development and displacement discourse. Pro-change and anti-change groups have hence emerged, fragmenting communities. There is also widespread criminalization of dissent in all the three cases. False and fabricated charges have been made against those protesting against the activities of large industrial houses and this has been consistent across the Niyamgiri/Vedanta, Posco and Kalinganagar debates. Furthermore, forms of dissent, largely non-violent, have resulted in large crackdowns, causing injuries and deaths of over a dozen people, mostly *adivasis*.

The Dongria Kondh tribe has received a great deal of national and international attention, which has contributed largely to not only how they (indigenous communities) are perceived, but has also affected policy processes in their favor. Unfortunately, the Kalinganagar debate has received the least media attention, and despite calls from Amnesty International to adequately address the 2006 killings of *adivasis*, voices of dissent have both been suppressed and criminalized. A closer look at the http://www.kalinganagar.com website reflects criminalization of the anti-Posco group Visthapan Virodhi Jan Manch (VVJM) as well the Posco Pratirodh Sangram Samiti (PPSS), stating that "local people were never against the idea of an industrial complex at Kalinganagar" and that "they were led astray by activists" from such organizations. 22,000 people in Dhinkia and three other *panchayats* did not want to sacrifice their traditional livelihoods to make way for the proposed Posco site at Jagatsinghpur (Bera, 2011a). Sayantan Bera (2011a) points out that even those labeled pro-Posco are not happy having to leave their ancestral lands, but seem to have little choice. Voices of dissent are translating into discontentment and perhaps even loss of self.

The Internet has been of some benefit to the PPSS in the anti-Posco struggle and even more significant in the case of Niyamgiri, and it is worth discussing the role of the Internet in these struggles.

Role of the Internet in empowerment and environmental activism

The Internet has been a highly beneficial tool for disadvantaged and marginalized community members for spreading information through e-mail, web pages and social networking websites. The Internet, which is considerably cheaper and easily available, even in some remote areas, and is not mediated, provides a forum for exchange of information, for finding people and organizations with similar interests, and also for networking. Individuals and small organizations can break through national and international boundaries and make their voices heard across the world. Those fighting against environmental injustice do not need to go through the mainstream media to disseminate information (Kutner, 2008). The "unmediated nature of the Internet", as stated by Kutner (2008), "can work to the advantage of traditionally marginalized groups", helping to create a political space free of media gate-keeping and profit. Kutner further suggests that power "has been given to relatively powerless segments of society through the use of Internet-based technologies" just by giving these groups the opportunity to "create and disseminate information".

This information can be about a community's own problems or about issues grassroots organizations work with, thus providing a forum for exchange of useful information that could be beneficial to the organization's cause. Policy information and citizen's rights are also available to those who would like to use such information toward getting environmental justice. Availability of information over the Internet about a particular issue or problem can in turn help create public opinion across national and international borders.

In the Niyamgiri case, the Internet has been useful in awareness, advocacy and networking. Websites such as http://www.niyamgiri.net provided a lot of information on the Niyamgiri struggle. Also, the availability of videos on the Dongria Kondh tribe and their way of life (such as Survival International's documentary *Mine: the Story of a Sacred Mountain*) helps audiences relate to them and stresses the importance of protecting the tribe's habitat. The documentary shows the Dongria

Kondh and their ways of living in a highly positive light, focusing on the tribe's self-sufficiency, spirituality and oneness with the earth, while also highlighting the looming threat of Vedanta's plant and the tribe's resolve to ensure that Vedanta leaves their ancestral lands. There was also a page on Facebook that appealed to users to help in protecting the tribe. Amnesty International's report *Don't Mine Us Out of Existence,* on the Dongria Kondh, also provided an international realm of opinions on issues pertaining to the tribe and human rights injustices by Vedanta. This was accompanied by videos uploaded on YouTube showing Amnesty activists demonstrating outside the Vedanta Resources office in London. There were also some discourses online and comparisons with the Hollywood movie, *Avatar,* directed by James Cameron, largely owing to Survival International's campaigns that compared the Dongria Kondh to the Na'vi tribe in the movie. *Avatar's* narrative draws on colonization allegory, and the Na'vi tribe's sustainable ways of living are threatened by the RDA Corporation in the film; the corporation itself was a well-grounded comparison to Vedanta and the Na'vi very reminiscent of the Dongria Kondh (Mishra, 2011).

An overview of Internet content shows that in the case of the Dongria Kondh, the narratives were bent on their spirituality, sustainable ways of living, their unique culture, self-sustenance, their connection to and dependency on their mountain, and lastly on the inherent need for protection of the tribe and its sacred mountain. The Dongria Kondh, through these narratives and discourses, became a metaphor for a pure and untouched natural world, fighting against the forces of modernization, manifested in the form of Vedanta.

The anti-Posco struggles in Jagatsinghpur and particularly the Kalinganagar issues have not attracted as much attention as Niyamgiri, and therefore there is not as much coverage on the Internet. Neither of these has had any international-level documentaries made about them. The Posco issue was more read and heard about widely, thanks to the work of the organization, PPSS, which even circulated its press release and letters to the Minister for Environment, Jairam Ramesh, online (Kafila, 2010) and also published appeals online (http://indigenouspeoplesissues.com). The organization has also published their appeal on Facebook.

The flow of information and communication between these grassroots organizations and local NGOs is worth noticing, with PPSS and VVJM (Kalinganagar) even coming together to share commonalities in their struggles against corporations and displacement. Networks have been built between these and other similar local and national organizations, creating a web of horizontal communication. Similarly, there has been a flow of information upwards from local to national and international levels, thus affecting knowledge and understanding, and inherently helping create a wider solidarity toward their cause.

Conclusion: Pluralism as environmental philosophy

Despite Orissa's fragile environment, the neoliberal reforms in the 1990s saw the government's focus shifting to rapid industrialization through mineral resource extraction, leading to the opening of the state's rich natural resources to exploitation by both national and international industrial houses. Orissa is at the heart of the environment crisis, being one of the poorest states, desiring to compete with the other states and to improve infrastructure and quality of life. Yet the state has to face an array of environmental disasters, most of which affect the poor. The debate on the environment and the rights of tribal peoples has been heavily contrasted with the intense desire of the Orissa government to develop. That explains the government's urges to tribal people in many parts of Orissa to accept the rehabilitation packages in order to allow for the construction of steel plants, opening of mines and other industrial projects. This has been a step grossly neglecting the needs of the environment and the interests of the tribal and other communities who are a part of that environment and are dependent on it. These tribal peoples, ironically, are not the ones who pose a threat to the environment, but instead are at the receiving end of all environmental conflicts.

There is no doubt, from the examples in this chapter, that people can stand up to corporations and affect the course of environmental justice. In Kalinganagar, despite Tata Steel's persistent efforts at wooing people through large-scale welfare activities and Self-Help Groups (SHGs), the people have been resilient and their struggles have continued. The anti-Posco struggles have also continued, and where forced acquisitions have occurred, there is great discontentment. The struggles of the Dongria Kondh against destruction of their sacred lands, and against forced physical, cultural, social and spiritual displacement are demonstrative of the power of people over corporations.

The displacement of indigenous people from their traditional homes, located in mineral-rich areas of the state, has meant that the protectors of the land have been replaced by those who would go on to cause the contamination of water bodies, deforestation and innumerable ecological destruction to these areas. At the heart of all tribal peoples' debates is the environment, as their ways of living are rooted in traditional sustainability. These voices are calling out for two things: protection of their environment and protection of their communities, culture and livelihoods attached to that very environment.

What is clear is that there are alternative voices that demand that they be heard, that their problems and issues be addressed. There is persistence in their struggles, and these struggles have drawn in people from local, national and international realms, creating networks of people that have the power to affect change.

This chapter hence calls for pluralism: the respect for life in all its forms; the respect for peoples and their human rights; respect for diversity and multiplicity of cultures, thoughts and ways of living. The concept of pluralism entails recognizing the needs of tribal communities themselves and hence does not lay down an imposition of mainstream political, economic and social motives.

I would like to extend pluralism to environmental discourse as well. Environmental movements have indeed shaped our understanding of the Earth in various political and social dimensions. As explored earlier in this chapter, indigenous peoples have their own notions of their environment, and as they live and depend on it, they respect and treat the rivers, hills, mountains and the various creatures that live in it with respect and reverence. Most indigenous philosophies dwell on the oneness of all living things; human beings just being a part of the web of life and not above it. We have something to learn from them; in order to protect our planet from the destruction caused by the patterns of our ways of living – excessive production and consumption of things we may not even need, and excessive waste that has no way of disappearing from the earth – we need to rethink and give way to alternative lifestyles and alternative thinking.

The discourses around protecting the planet have continued but so far they have not come up with solutions to our environmental problems. These indigenous people, who may not possess all the things we do, however, know more about the Earth: how to take just what is needed from the Earth's resources; how to protect and respect the trees and rivers for the future and so on. That is what we can learn from them – by letting them choose what kind of life they would like to live.

Lastly, the chapter drawing on the discourses on Orissa's environment and development calls for a rethinking of development based on extraction of natural resources. Sadly, climate change is not avoidable, and it is not possible to continue exploiting the Earth without any further consequences. Orissa's climate has undergone rapid changes and a climate catastrophe may not be far away. The ones that have the most to suffer from such climatic changes have been and will continue to be the poor and the vulnerable. Therefore, there is a need for focus on environment and climate change mitigation in policy frameworks and practice, rather than blind economic growth that has no regard for the environment.

The philosophies of the Native Americans, perhaps among all others, display a plethora of wisdom on the Earth and the relationship between humans and their environment. These bits of wisdom can serve as a reminder for immediate action and a rethinking of political, economic and environmental ethics, policies and practices:

> Only when the last tree has died,
> And the last river has been poisoned,
> And the last fish has been caught,
> Will we realize that we cannot eat money.
>
> –Cree Proverb

Notes

1. The author would like to express her gratitude to her father, Mahesh Chandra Mishra, for his constant support and valuable advice provided regarding this chapter.
2. Taken from Chief Seattle's speech cited in Kerry-Ward (2009).
3. In 2011, the State of Orissa was renamed to "Odisha".
4. Gram sabha is a meeting of all adults at the village level.
5. Their demands also include the following: (i) rights over acquired land; (ii) action against officials "responsible for the killings" and (iii) a ban on multinational entry (Pandey, 2006).
6. Village-/town-level local governments.
7. Dhinkia is a village in Jagatsinghpur that falls under the proposed Posco site.

References

Agarwal, A., Sharma, A., & Narain, S. (2001). *Poles Apart: Global Environmental Negotiations 2*. New Delhi: Centre for Science and Environment.

Amnesty International (2010a). *Don't Mine Us Out of Existence: Bauxite Mines and Refinery Devastate Lives in India*. London: Amnesty International Publications. Available at: http://www.amnesty.org.uk/uploads/documents/doc_20144.pdf. Accessed on 10 October 2010.

Amnesty International (2010b). *Attacks on Indigenous Protestors in Orissa Continue*. Available at: http://www.amnesty.org.nz/news/attacks-indigenous-protestors-orissa-continue. Accessed on 5 December 2012.

Amnesty International (2012a). *Vedanta's Perspective Uncovered: Policies Cannot Mask Practices in Orissa*. London: Amnesty International. Available at: www.amnesty.org/en/library/info/ASA20/029/2012/en. Accessed on 12 December 2012.

Amnesty International (2012b). *India: Vedanta's Human Rights Promises "meaningless and hollow"*. Available at: http://www.amnesty.org/en/news/india-vedantas-human-rights-promises-meaningless-and-hollow-2012-08-28. Accessed on 12 December 2012.

Amnesty International (2012c). *India: Authorities Must Ensure Justice for Victims of Kalinganagar Shootings Six Years Ago*. Available at: http://www.amnesty.org/fr/library/asset/ASA20/001/2012/en/f2736735-91ea-4594-b20e-039fd06ae2a9/asa200012012en.pdf. Accessed on 5 December 2012.

Amnesty USA (2007). *Kalinga Nagar Police Firing One Year On – Orissa Must Ensure Speedy Justice for Adivasi Victims and Address their Concerns Over Displacement*. Available at: http://www.amnesty.org/en/library/info/ASA20/001/2007/en. Accessed on 14 October 2010.

Baboo, B. (1992). *Technology and Social Transformation: The Case of the Hirakud Multi-purpose Dam Project in Orissa*. New Delhi: Concept Publishing Company.

Baboo, B. (1997). Big Dams and the Tribal: The Case of the Hirakud Dam Oustees in Orissa. In G. Pfeffer, & D. K. Behera (Eds.), *Contemporary Society: Tribal Studies (Development Issues, Transition and Change)* (pp. 92–104). New Delhi: Concept Publishing Company.

Bera, S. (2011a, 11th February). POSCO Unplugged. *Down to Earth*. Available at: http://www.downtoearth.org.in/content/posco-unplugged. Accessed on 5 December 2012.

Bera, S. (2011b, 16th December). Violence at POSCO Site. *Down to Earth*. Available at: http://www.downtoearth.org.in/content/violence-posco-site. Accessed on 5 December 2012.

Campaign for Survival and Dignity (2010, 15th May). Timeline of Events in Posco Project Area Relating to Forest Rights. Available at: http://www.countercurrents.org/fra150510A.htm. Accessed on 2 October 2010.

Down to Earth (2008, 31st January). Revisiting Kalinganagar. Available at: http://119.82.71.32/full6.asp?foldername=20080131&filename=news&sec_id=50&sid=38. Accessed on 18 October 2010.

Forest Survey of India (2009). India State of Forest Report 2009. Ministry of Environment and Forests, Government of India. Available at: http://www.fsi.org.in/sfr2009/orissa.pdf. Accessed on 11 October 2010.

Forest Survey of India (2011). India State of Forest Report 2011. Ministry of Environment and Forests, Government of India. Available at: http://www.fsi.org.in/final_2011.pdf. Accessed on 8 December 2012.

Ghosh, P. (2010). Fresh Trouble for Posco's Orissa Plant. Available at: http://www.livemint.com/2010/05/26223645/Fresh-trouble-for-Posco8217.html. Accessed on 17 October 2010.

Government of India (2006). *Forest Rights Act*. Delhi: Government of India.

Government of Orissa (2005). *Orissa: Minerals of Orissa*. Available at: http://india.gov.in/knowindia/mineral_orissa.php. Accessed on 10 October 2010.

Government of Orissa (2008). *Report On the Activities of Steel and Mines Department For the Year 2007–08*. Available at: http://www.orissaminerals.gov.in/Download/Annual_Report.pdf. Accessed on 5 May 2013.

Government of Orissa (2010). *Orissa Climate Change Action Plan*. Available at: http://orissa.gov.in/portal/occap.pdf. Accessed on 11 November 2010.

Hindustan Times (2010). Posco Violated Forest Rights Act, Confirms Panel. 3rd November. Available at: http://www.hindustantimes.com/Posco-violated-forest-rights-act-confirms-panel/Article1-621891.aspx. Accessed on 8 November 2010.

Indigenous Peoples Issues (2010). Appeal from POSCO Pratirodh Sangram Samti (PPSS): Opposed To Proposed POSCO Steel Plant Project. Available at: http://indigenouspeoplesissues.com/index.php?option=com_content&view=article&id=5503:appeal-from-posco-pratirodh-sangram-samti-ppss-opposed-to-proposed-posco-steel-plant-project&catid=63:central-asia-indigenous-peoples&Itemid=85. Accessed on 4 November 2010.

Kafila (2010). Posco Pratirodh Sangram Samiti to Jairam Ramesh. Available at: http://kafila.org/2010/08/19/posco-pratirodh-sangram-samiti-to-jairam-ramesh/. Accessed on 3 November 2010.

Kerry-Ward, A. (2009). *The 20th Century and Then What: A Philosophical View of Life*. Leicester: Troubador Publishing Ltd.

Kutner, L. A. (2008). Environmental Activism and the Internet. In K. R. Gupta, M. A. Jankowska, & P. Maiti (Eds.), *Global Environment: Problems and Policies Vol II* (pp. 181–190). New Delhi: Atlantic Publishers and Distributors.

Mahapatra, R. (2006). Disaster Dossier: The Impact of Climate Change on Orissa. *Infochange*. Available at: http://infochangeindia.org/200603265760/Environment/Features/Disaster-dossier-The-impact-of-climate-change-on-Orissa.html. Accessed on 2 October 2010.

Manecksha, F. (2012, 22nd June). Against all Odds, a Struggle Continues. *Times of India*. Available at: http://www.thehindu.com/opinion/op-ed/against-all-odds-a-struggle-continues/article3555520.ece. Accessed on 10 December 2012.

Milstein, T. (2009). Environmental Communication Theories. In S. W. Littlejohn & K.A. Foss (Eds.), *Encyclopedia of Communication Theory* (pp. 344–349). Thousand Oaks, CA: Sage.

Mishra, M. (2008). *How Have the Media Addressed Rice-fish Farming in Coastal Orissa, India?* Reading: University of Reading (MSc Dissertation).

Mishra, M. (2011, June). *Environmental Communication in India: Lessons from Orissa*. Paper presented at AMIC Annual Conference on Taking Stock of Media and Communication Studies: The Challenges and Opportunities of Globalisation, New Media and the Rise of Asia, Hyderabad, India.

Mukherjee, K. (2010, 9th August). POSCO's Orissa Steel Mill May be Cleared After Aug 16. Reuters. Available at:http://in.reuters.com/article/idINIndia-50719120100809. Accessed on 15 October 2010.

NDTV (2010, 13th October). Govt Identifies over 10,000 Villages as Drought-hit in Orissa. Available at: http://www.ndtv.com/article/india/govt-identifies-over-10-000-villages-as-drought-hit-in-orissa-59314. Accessed on 16 October 2010.

Outlook India (2010). 85,000 People Affected in Orissa Floods. Available at: http://news.outlookindia.com/item.aspx?689846. Accessed on 12 October 2010.

Pandey, B. (2008). The Kalinganagar Tragedy: Development Goal or Development Malaise. *Social Change*, 38(4), 609–626.

Patnaik, N. (2010, 22nd July). Drought Year Stares Orissa as Monsoon Plays Truant. *The Economic Times*. Available at: http://economictimes.indiatimes.com/

news/politics/nation/Drought-year-stares-Orissa-as-monsoon-plays-truant/articleshow/6202536.cms. Accessed on 16 October 2010.

Pradhan, P. (2006). Police Firing at Kalinganagar. People's Union for Civil Liberties (PUCL) Bulletin. April 2006. Available at: http://www.pucl.org/Topics/Dalit-tribal/2006/kalinganagar.htm. Accessed on 15 October, 2010.

Press Trust of India (2012, 5th December). *Vedanta's Alumina Refinery Shuts Down in Orissa*. Available at: http://www.ptinews.com/news/3187115_Vedanta–s-alumina-refinery-shuts-down-in-Odisha-. Accessed on 8 December 2012.

Rehman, M. (2010, 24 August). India Blocks Vedanta Mine on Dongria-Kondh Tribe's Sacred Hill. *The Guardian*. Available at: http://www.guardian.co.uk/business/2010/aug/24/vedanta-mining-industry-india. Accessed on 5 September 2010.

Reserve Bank of India (2010). *Handbook of Statistics on the Indian Economy*. Mumbai: Reserve Bank of India.

Satapathy, D. (2012, 11th December). Can Vedanta restart Lanjigarh Refinery. *Business Standard*. Available at: http://www.business-standard.com/taketwo/news/can-%20vedanta-restart-lanjigarh-refinery/495227/. Accessed on 12 December 2012.

Shiva, V. (2002). *Water Wars: Privatization, Pollution and Profit*. Cambridge, MA: South End Press.

Survival International (2010a). Troubled Vedanta Loses Appeal for Controversial Refinery 21 October. Available at: http://www.survivalinternational.org/news/6605. Accessed on 10 October 2010.

Survival International (2010b). Vedanta's India Mine Slammed in Devastating Government Report 16 August. Available at: http://www.survivalinternational.org/news/6368. Accessed on 10 October 2010.

The Economic Times (2012, 21st April). Tata Steel to Commission Kalinga Nagar Plant by March 2013. Available at: http://articles.economictimes.indiatimes.com/2012-04-21/news/31378996_1_kalinga-nagar-jamshedpur-expansion-director-h-m-nerurkar. Accessed on 3 December 2012.

The Hindu (2010a, 3rd April). Tension Continues in Kalinganagar. Available at: http://www.hindu.com/2010/04/03/stories/2010040351970300.htm. Accessed on 15 October 2010.

The Hindu (2010b, 29th May). Orissa Govt. Rejects Demand to Shift Posco Plant Site. Available at: http://www.thehindu.com/news/national/article441478.ece. Accessed on 20 October 2010.

The Hindu (2012, 9th November). *Tata Steel slips into loss in Q2*. Available at: http://www.thehindu.com/business/companies/tata-steel-slips-into-loss-in-q2/article4082031.ece. Accessed on 10 December 2012.

Times of India (2006, 3rd January). 13 die in Kalinga Nagar violence in Orissa. Available at: http://timesofindia.indiatimes.com/india/13-die-in-Kalinga-Nagar-violence-in-Orissa/articleshow/1356534.cms. Accessed on 15 October 2010.

Times of India (2010, 21st October). Another Setback for Vedanta, Alumina Refinery Plan Fails to Get Nod. Available at: http://timesofindia.indiatimes.com/business/india-business/Another-setback-for-Vedanta-alumina-refinery-plan-fails-to-get-nod/articleshow/6788195.cms. Accessed on 2 November 2010.

Wong, L. (2008). Resisting and Transforming Space: The State, Environment and Indigenous People in Malaysia. In K. R. Gupta, M. A. Jankowska, K. Bosselmann, & P. Maiti (Eds), *Global Environment: Problems and Policies, Volume III* (pp. 146–173). New Delhi: Atlantic Publishers and Distributors.

Xaxa, V. (2012). Identity, Power and Development: The Kondhs in Orissa, India. In S. Sawyer & E. T. Gomez (Eds), *The Politics of Resource Extraction: Indigenous Peoples, Multinational Corporations and the State* (pp. 180–203). Basingstoke, UK: Palgrave Macmillan.

8

Socio-cultural Perspectives on Sustainability of Sexually Transmitted Diseases, HIV and Pregnancy Prevention among Thai Students in Bangkok

Patchanee Malikhao

Introduction

Globalization is a discourse. It depicts how people from different states, nations or territories are becoming more interconnected and interdependent on one another. Manifestations of globalization exist in different arenas: economic, political, cultural, technological, social and ideological. Globalization involves development, progress and disparity. For some, it may be the future of opportunities and wealth but for others it may cause unequal access to information, opportunities and wealth that leads to economic distress and unhappiness. It cannot be denied that communication technology in the dominance of capitalism, influenced by neoliberal ideologies, is the propelling factor of globalization so that there exist worldwide exchanges in labor, trade, technology, production and capital.

Zooming in on globalization

Globalization is not just what is happening in this century – that is, the reduction of barriers physically, legally, linguistically, culturally and psychologically – so people can engage with one another no matter where they are (Scholte 2005: 59). In fact, many history scholars reveal that globalization has occurred since the sixteenth century, in the forms of cultural exchanges via bilateral and, later, transnational connections.

According to Hopkins (2002: 1–10), globalization is a new terminology that denotes the following ongoing historical process: first, archaic globalization; second, proto-globalization; third, globalization; and fourth, post-colonial globalization.

Hopkins explains further that archaic globalization occurred from Byzantium and Tang to the renewed expansionism of Islamic and Christian power after the 1500s. He identified proto-globalization with the political and economic developments that became especially prominent between about 1600 and 1800 in Europe, Asia and parts of Africa. The third historical process, *globalization*, he refers to as the colonial period from the 1760s onwards. Globalization that can be related to modernity started from 1800, according to Hopkins. It refers to the rise of the nation state and the spread of industrialization. The last process, post-colonial globalization, refers to the contemporary form that can be dated approximately from the 1950s.

Globalization has brought changes in the way people think, behave and pass information on to later generations, in other words: culture. With help of modern communication technology, the transfer of cultural products via the mass media has been more pronounced and one can see the fluidity of localized culture aided by communication.

Sexual culture, HIV/AIDS and sustainable development

Changes in sexuality have been documented in different phases of globalization, as I will elaborate later. Sexuality, according to Hollander, Renfrow and Howard (2011: 20), refers to a group of concepts: eroticism, sexual behaviour, sexual orientation and desire to engage in sexual activity, which is closely related to gender. A local culture has had an enormous impact on the regulations and expressions of an individual sexuality. Not only has a local culture evolved through phases of globalization, but sexuality also has. Public health policy makers need to keep pace with the current sexual culture that evolves with new communication modes and media to prevent and campaign against one of the social problems in the contemporary globalized world: the HIV/AIDS issue. HIV infection is related to social disintegration, gender relations, income distribution, power, and lack of awareness and information from well-planned health communication programs either from the government or from the community. HIV/AIDS is thus not just an individual problem.

HIV/AIDS is a threat to sustainable development in many developing countries such as those in Africa and South-East Asia, where the working population is not healthy enough to work and the cost of anti-retroviral therapy per person is so high that not everyone can get access to it. Malikhao (2012: 3–5) explains that HIV/AIDS goes hand-in-hand with poverty and social injustice. Gender inequality as a result of globalization phases in each country and marginalization are at the root cause of HIV/AIDS epidemics. HIV/AIDS prevalence among drug users and heterosexual partners are high among those in developing countries. Indeed, a country faced with a HIV/AIDS epidemic is far away from sustainable development. Now, we come to a new term: sustainable development.

Jabareen (2011: 388–392) proposed a theoretical framework for sustainable development consisting of ten concepts: the compromising between the domination of nature and the intrinsic right of nature; keeping the natural capital constant; envisioning and planning alternative urban life; equity of power and resource distribution; uncertainty management in the light of climate change; efficient urban planning that could lessen global warming by way of using ecological friendly materials and space management among buildings and traffic ways to lessen traffic congestion and reduce toxic exhausting gas to the environment; the integrative concept of environmental, social and economic concerns in planning and management; use clean, renewable and efficient use of energy; emphasis on environmentally sound economics; and politics of sustainability, which goes beyond environmental issues such as security, peace, trade, hunger, shelter and other basic services.

From this definition, it is clear that HIV/AIDS is tied with sociocultural and economic concerns in the tenth category of politics of sustainability that goes beyond environmental issues only.

HIV/AIDS in Thailand: Root causes and prevention strategies

Thailand is an interesting case study for HIV/AIDS because it has different sexual norms and values to the West, such as socially accepted practices of commercial sex as part of its hierarchical and patriarchal culture (Klausner 1997). It has never been colonized, yet it has been "absorbing" Western ideas, technologies and values since the reign of King Chulalongkorn (1868–1910) (Stifel 1976: 1184). Thailand is now part of an informal United States empire in the contemporary globalization period (1954–present) (Malikhao 2007: 74). With a long history of female subordination through polygamous sexual behavior together with the industrialization via imported Western technology and know-how, females are the victims of sex trade and sex tourism

(Malikhao 2007: 65–77). One of the factors that help spread HIV is the hybridized sexual culture as influenced by the pop culture. Malikhao (2007: 77–78) observes that Thailand has experienced a *hybridization of sexuality* – such as scoring for sexual encounters for both males and females, trading sex with expensive accessories, males who have sex with males and so on – since the start of the contemporary globalization period.

According to the Thailand-United Nations Assembly Special Session (UNGASS) Country Progress Report (2008: 10), the number of new HIV infection cases was estimated to be 13,936 in 2007. The projection of prevalence of HIV from 2008 to 2011 is high among females who got it from their husbands, males who have sex with males, males from sex workers, males from wives, extra-marital sex and injecting drug users, respectively (UNGASS 2008: 32). It is alarming that the percentage of young women and men aged 15–24 who are HIV infected is now 0.64 per cent compared, to 0.45 per cent in 2006 (UNGASS 2008: 21).

Much interest has been given to HIV prevention instead of spending a lot of money to treat those who are already affected by the epidemic. Scholars identify three sorts of prevention: biomedical, behavior change and structural. Biomedical prevention includes the use of male and female condoms, cervical barriers, male circumcision, curable sexually transmitted infection control, anti-retroviral therapy, and control of mother-to-child transmission. HIV prevention by emphasizing behavioral change includes client-initiated HIV counseling and testing, treatment for drug addictions, and school-based intervention. These two types of prevention are not enough if the policies and laws do not support behavior change. Therefore, structural changes need to include a range of interventions, such as changes in laws, prices and/or taxes, subsidies, vouchers, housing, income-generating activities, women's empowerment and so on (Galarraga et al. 2009: 6).

Thailand has been combating the HIV/AIDS epidemic since the first case of AIDS in Thailand caught the media's attention in 1984 (UNDP 2004:7). The initial attempt to silence the importance of the pandemic and the quarantaining of people in the late 1980s was followed by a series of HIV/AIDS prevention approaches which combined media advocacy and social marketing with biomedical prevention in the late 1990s. Massive television campaigns nationwide and a countrywide "100% Condom Program" started in 1992, when the Minister of Public Health gave away 60 million condoms for free (UNDP 2004: 17–18).

That was a successful program, but after the Asian Economic Crisis in 1997, the government had to reduce its campaign budget by 28 per cent (D'Agnes 2001: 350). The government then turned to support the existing and less expensive non-governmental organization (NGO) HIV/AIDS *community-based approach* (UNDP 2004: 40). This can be considered part of a multi-sectoral HIV/AIDS prevention which includes the collaboration of the government, NGOs and the private sector for both clinical and social interventions. The Thai government also supports interfaith-networking approaches to prevent HIV by emphasizing the ABC notion (Inter-Faith Networking on AIDS 2004: 2–3). I studied the *community faith-based strategies*, both Buddhist and Christian, in the North of Thailand in 2005. They are a mixture of biomedical prevention, faith-based intervention and civic organization, as further detailed in my book *Sex in the Village: Culture, Religion and HIV/AIDS in Thailand* (Malikhao 2012).

In the 2000s, there was a cry for sex education in schools that support alternatives for safer sex (http://www.manager.co.th, 10th February, 2007, accessed on 12th February, 2007). *School-based interventions* became the new slogan as they could introduce wide-scale behavior change, attempting to reduce HIV risk-taking behavior. As a result, Thailand launched sex education in pilot schools in 2004 and has covered 98.7 per cent of the provinces in Thailand since (http://teenpath.net/teenpath/database/Coverage-Province.asp, accessed 22nd October, 2009). This includes sex education in middle schools, vocational schools and high schools. Today there are 62 schools in Bangkok that teach sex education (http://teenpath.net/teenpath/ database/Summary.html, accessed 22nd October, 2009).

According to the UNGASS report (2008: 41), sex education was supposed to be taught in secondary and vocational schools during 16 periods per academic year. The curriculum prescribed that Thai students should understand the importance of gender relations, rights, reproductive systems, sexuality, HIV and sexually transmitted diseases, and prevention. However, the Thai national sexual behavioral survey of 2006 reports that only 39.5 per cent of male and female youths aged 15–24, answered five basic knowledge questions about HIV correctly (47 per cent of these were males and 2.3 per cent were females) (UNGASS 2008: 40).

The latest 2006 national AIDS prevention plan, as reported by the Ministry of Public Health on their website (http://www.aidsthai.org), includes communication for teenagers to prevent HIV. The "sex education in school" curriculum is being posted for public comments on the website of *teenpath*, launched by the Ministry of Education with financial support from the Canada Fund, the Global Fund to Fight AIDS, Tuberculosis and Malaria and the American Non-Profit

Organization PATH (see http://www.path.org and http://www.teenpath.net). The content emphasizes safer sex, gender relations, abstinence and dialogs.

HIV/AIDS prevalence is a manifestation of unequal development. Globalization spurs hybridized sexuality and that influences the HIV/AIDS prevention strategies. To achieve sustainability in development, one needs to keep pace with new kinds of sexuality and sexual values the locals consume by copying the global at a high speed.

This study was designed to assess HIV prevention among Thai youth and the factors that influence their sexuality, gender equity, sexual and reproductive health and rights in view of a sustainable future. With that in mind, this author aimed to conduct a qualitative research project to gain in-depth information from the so-called generation Y so that the knowledge gained will be useful to help update the current HIV/AIDS prevention policies. Hence, the study, undertaken in Thailand during July and August of 2010, included focus group discussions and in-depth interviews. The venues of study were two co-ed high schools right in the hub of Bangkok: a private one and public one; and three different universities: one private university, one state university and a former teachers' college. Students from the private high school represent a middle-class background, while those from the state high school a lower-income background. It is assumed in this study that the students from the teacher's college come from a lower-middle-class background, those from a state university the middle class, and the private university represent the upper-middle class.

The rationale for selecting ninth-graders was because this age group was one of the first that received the basic sex education curriculum on gender equity and safer sex practice. The high school seniors will receive a more comprehensive sex education training according to their age group. As the sex education program has been implemented six years ago, students who are now in college must have passed the sex education exam in school already. I tried to evaluate the sexuality trend, HIV prevention, sexual rights and media use of the students under study.

Specific aims of this research

- To explore sexuality trends of Thai youths aged 15–24, from an upper-middle-class and a working-class background.
- To investigate the socio-economic and/or cultural impacts on sexuality, HIV and sexually transmitted disease (STD) prevention and

sexual rights of students from upper-middle-class and working-class background in Bangkok.

- To understand media habits for print media, electronic media (TV, radio) and new media (e-mails, websites, social networking websites) of the people under study.

Theoretical perspectives

I explored the interrelationship of the following broad concepts: globalization, the mass media, Thai culture, sexuality and HIV/AIDS.

Malikhao (2007; 65–77) explains that *Thai culture* has two distinguishing features: patriarchy and the Buddhist worldview that reinforces female subordination. Patriarchy emerging in the early globalization period and the pre-marital sex value introduced in the globalization period gave rise to hybridized sexuality such as prostitution. This factor can be seen as endogenous because it has been part of Thai culture for centuries. In the contemporary globalization period, more exogenous factors, such as the arrival of the Americans before and during the Vietnam War, the capitalistic economy, the import of Western communication technology and an economy based on neoliberal principles, have contributed to the growth of the sex industry and diversified sexuality. The exogenous factors are products of *cultural globalization*, which can create borderless interconnections but can also initiate a separate identity among local recipients, such as currently localized or hybridized forms of Thai sexuality.

In Malikhao (2007: 78–80), I explained that Thai sexuality in this globalization period can be categorized as traditional, convergent, divergent, and hybridized. Traditional sexuality includes sex outside of marriage, that is, sex with mistresses and sex in brothels. Convergent (globalized) sexuality is seen in the greater visibility of homosexuality; sexploitation of females and males from other countries; a tendency for sex with girlfriends rather than with sex workers for adolescents; and a tendency toward pre-marital sex among adolescent females. Divergent (local globalization) forms of sexuality find expression in the export of Thai sex workers to Europe and Japan. Hybridized forms of sexuality take a number of forms: the sale of one's virginity; charity sex, meaning selling sex in exchange for as little as a fashionable outfit, a mobile phone and so on; phone sex; sex in massage parlors, in karaoke bars, or women's dormitories; the posting of nude photos onto websites; the first sexual encounter on Valentine's Day; or so-called sex "races", where competing for casual partners has become popular and lists and ranking of "hits" are shared and shown around as badges of honor.

In this contemporary globalization period, Thais experience the symbolic representation of a post-modern "culture industry", where people are introduced to symbolization of the capitalistic market economy via advertisements and images portrayed in films and movies (Tomlinson 2001 19). Taylor (1997: 20) explains that post-modern cultural forms are more concerned with form rather than content and that all have become commodified. That is due to the mass media.

"Globalization" is about the emergence of completely new social, political and business models that interlink governments and big businesses and that have a great impact on every aspect of society (Friedman 2007: 48). Nash (2000: 53) explains how the mass media – for specific definitions of "old" and "new" mass media, see McQuail (2005: 51–52) – play an important role in the contemporary globalization process. He states that mass media help create a global consciousness, by which people can compare how others live with their own local living conditions. Thompson (1995) states that, although the mass media diffuse their messages globally, consumers of the mass media consume the messages locally. This creates a process that he calls *the axis of globalized diffusion and localized appropriation*. Thus, the mass media (and communication in general) are inevitable and crucial in the globalization process. Elliott and Lemert (2006) propose that globalization has a profound impact on the individual level, and that causes a new kind of individualism. They define the new individualism as a highly risk-taking, experimental and self-expressing individual underpinned by new forms of apprehension, anguish and anxiety. As each individual has become a consumer of the media conglomerate in a capitalist society, the impact on each person's experiences of identity, gender, sexuality and family life is the topic discussed by many scholars. Elliot and Lemert (2006: 114) note that sexuality in the United States is currently framed and regulated through mass media, advertising and information culture as a consequence of globalization and that various forms of sexuality can be called "discursive sexuality" among the new generation. Among American youth culture, there is evidence that increasing globalization within media systems in the United States has shaped the degree of individualism in society (Elliot & Lemert 2006: 4–5) and contributes to new attitudes toward sex, sexuality and individual identity. Taylor (1997: 20) explains that post-modern cultural forms are more concerned with forms rather than content and have become commodified. High levels of individualism can lead to narcissism. Twenge and Campbell (2009: 19) state in their book, "The Narcissism Epidemic", that the central feature of narcissism is a very positive and inflated view of self and this value is growing rapidly in the American culture fuelled by mass media, including new media, and changes in parental approaches to upbringing that emphasize self-expression. Symbolic representations of the new American culture of self-expression are the emphasis on celebrities in the media, the success of MySpace and Facebook as social networking sites, the uploading of personal videos on YouTube, twitter (micro-blogging and text-based social networking or SMS on the internet via its own website) and blogging (Twenge and Campbell 2009).

Hassan (2008: 27) states that globalization fueled by the mass media and information technology advocates a "vast expression of access to information, the centrality of the internet, and networked communities" while at the same time the information received can be meaningless; the loss of real community is evidenced; the lack of time for reflection can cause superficial and hurried cultural forms. Moreover, here arises a new problem: the so-called digital divide engenders an information gap between those who can afford the hardware and software and access the Internet and those who cannot afford such access to information (Scholte 2005: 36).

In other words, I was interested to study whether Thai youths are influenced by mainly Americanized media and display similar characteristics as American youth, and whether sex education at school has helped them with their sexuality, HIV and STD prevention and sexual rights.

Research questions for the study

- How do the Thai research participants aged 15–24 from an upper-middle-class or a working-class background under study perceive sex education taught as related to their sexuality, HIV and STD prevention, and sexual rights?
- Which socio-economic and/or cultural factors do the Thai participants under study, who are from an upper-middle-class versus a working-class background, think hinder or advance their access to sexual and reproductive health, including HIV and STD prevention?
- What are the differences and/or similarities in the sexuality, HIV and STD prevention, and the rights of the male and female youths under study who are from an upper-middle-class or a working-class background?
- How has the mass media and the new media influenced the participants from an upper-middle-class versus a working-class background on their self-identity, sexuality and sexual rights?

Research design

Research sites: A private and a public secondary school in the center of Bangkok; a private university, a state university, and a former teachers' college which has been upgraded to become a university, all of which are situated in Bangkok and Pathum Thani.

Data gathering method

Focus groups of male and female students were conducted. They are the ninth- and twelfth-graders of high schools and college students. Each focus group consisted of five to seven people. Around 20 high-school students and 10 college students were invited for additional in-depth interviews. School councilors and sex education teachers were also invited for in-depth interviews. Sex education textbooks were analyzed. Data were collected during July and August 2010.

Findings

From an analysis of the focus group discussions, it became clear that both male and female adolescents' attitudes toward the role of women and men in society have not changed from that in the past. Females are thought to have equal opportunities as males, as far as most occupations and education are concerned. A bit more probing is a controversial "occupation", such as being a female monk. It was unthinkable for the students under study to picture a female monk. Both male and female adolescents seem to accept the role of females as nuns who are of lower status and adhere to only eight precepts compared to monks who have to adhere to 227 precepts. Some male students mentioned that the nuns' duties as sweeping and mopping the floor in the temple. Most students do not know that female monks adhere to 311 precepts and that there are two female monks in Thailand. They did not seem to like the idea that females could attain the same status as male monks. "It is strange; it is not normal", they reported.

For household responsibilities, female adolescents are expected to help with the household chores more than the males. Most male students accept that males should take part in child-caring and raising and household responsibilities. Whether they actually do participate in household chores is another question.

Even though more than 50 per cent of the students interviewed stated that it is quite normal to live together without being married, they would prefer a couple to finish their studies and get a job first.

Should females have the same sexual rights as males?

Gender equality is still controversial when it comes to sexual values and rights. About half of the male students under study stated they would not mind if their girlfriends already had a sexual experience with someone else, while the majority of girls under study stated that virginity is still very important to them. The students interviewed think that being a male has more advantages than being a female, because a female can get pregnant. Some students mentioned that females are not wanted in certain societies such as China; some indicated that females wanted to be males in their next life. All students under study regard females who have many sexual partners worse than males.

Students interviewed don't seem to grasp the concept of *sexual harassment*, in the worst-case scenario: rape. Only one male student from a private college indicated that it was the men's fault only. Most students answered rapidly that a woman who wears too short a skirt or a tank top "invites" sexual harassment, even rape. For those who got raped but wore decent clothes, they explained that because the rapists got sexually aroused from those who wear sensual clothes so they raped any woman when an opportunity arises. When asked whether it is really women's fault, the students responded that women are 50 per cent responsible and men who are drug addicts or homeless are to be blamed for not being able to control themselves. Their answers contradict what is stated in the handbook of sex education for youth for grades seven to nine, published by PATH, which states:

> Societal values make sexual harassment just and have no mercy on the victims such as the beliefs that women who do not cover themselves properly intend to seduce men and those women deserve to be harmed. Wearing clothes is an individual rights but a person who could not control his sexual desire but blame the clothing of the counterpart as seductive is violating private rights because one cannot do whatever one wants without thinking of the consequence to others.
>
> (PATH 2007: 150)

Browsing in the textbooks students used in their sex education course, published by the Ministry of Education, I found that there are many sentences here and there which urge females to wear covered clothes because sensual clothes may bring harm, and that females should not go out at night and so on. This indicates the safety and patriarchal issues women are facing.

All male students know about *condom use*. Many of them indicated that it was embarrassing to go and buy a condom from a convenience store (such as a 7/11 store) especially when the cashier is a woman. They would prefer to see more condom vending machines in male bathrooms in department stores or malls. They disagreed with having a condom vending machine in schools because that would make younger students curious, and they would feel embarrassed that someone would see them taking a condom out of the machine. High school students suggest there should be a counseling room where students can go and talk about their problems and get condoms. College students also suggested that there should be a center that offers a sex education course or provides counseling services to them.

The students under study reported that they learned different birth control methods but they could not remember much in detail. Some did not even know the correct safe period before and after ovulation (the safe period does not mean that it is completely safe for unprotected sex). In-depth interviews with a girl who was pregnant in a high school and a young male college student whose girlfriend gave birth to a baby a year ago indicate that students think that the chance of getting pregnant is rare and they could not apply what they learned in school to be used in real life regarding safer sex and unexpected pregnancy prevention. It is strange to learn that most male students wanted women to be responsible for sterilization and birth control. A few male students at a public high school stated that a permanent birth control for men is the same as a castration. Most of them do not know that permanent birth control in men can be done easier than in women. Worst of all, all of the students and a young sex education teacher chosen to be interviewed do not know who Mr. Mechai Viravaidya is, the internationally renowned Thai founder of the Population & Community Development Association (PDA). Khun Mechai is also known as the *Condom King* for his famous condom promotion campaigns. As a result, older generations still humorously call condoms a "Mechai". This was not known to any of the students or the teacher. This can be considered a serious gap because these successful condom promotion campaigns were launched in the 1990s. People who were born later do not have a grasp of what has been achieved regarding population control or HIV/AIDS prevention campaigns in earlier time periods.

Most students report quite positively about their attitudes toward *homosexuality*. However, many students under study indicated that upbringing and environment forge homosexuality, and they think that homosexuality can be cured. Some students under study thought that homosexuality sits in the genes.

How did the sex education course help the students understand birth control, HIV/AIDS and STDs?

All students interviewed understand the importance of sex education. They envisioned it as a separate course from "Health Education". They suggested that it should start from year seven onwards. Some students even suggested having such a course at a primary school level. Students stated that sex education should concern relationships, sexuality, sexual rights and sexually transmitted diseases including HIV/AIDS as well as reproductive organs.

Regarding knowledge questions, it was found that students lack a good understanding of birth control, STDs and HIV/AIDS. Some students in a public school indicated that they already had sex with their girlfriends but they did not really know much about STD and HIV transmission. A few male and female students did not know that HIV is a virus. They had no idea what a virus and a bacterium was. When asked how they study, it was clear that some sessions of the curriculum were not offered and they were left to do their homework. Some teachers do not want to teach sex-education out of embarrassment. In-depth interviews with two male sex education teachers in a public high school indicate there is a conflict between groups of conservative versus liberal teachers. The conservative teachers prefer tough punishments if students arrive late at school while the liberal ones try to understand the context in which a student grew up. An in-depth interview with a counselor who was a sex education teacher in a private high school indicates that single female teachers feel teaching sex education is embarrassing, so they do not want to teach these topics. Worse than that, a conservative school board did not want a female teacher to teach sex education despite the fact that she was well trained from PATH. Most sex education teachers are male physical education teachers in both public and private high schools.

What did students think affected the teaching and learning about sex and safer sex?

The students under study reported that they would prefer a co-ed classroom and small group discussions lead by a teacher. They indicated that one-way communication and a textbook are really dry and not conducive to learning. College students under study indicated that sex education should be provided at college too as non-credit sessions. Both high school and college students need a trustworthy, friendly and helpful counselor whom they can meet in private and ask about personal relationships, problems at home or at school, birth control and STD or HIV prevention. From in-depth interviews with sex education teachers, social class drives students to have early sexual encounters. Students from the public high school are those who come from decaying neighborhoods. Some students take and sell drugs at school. Some students are exposed to the sexual behavior of their parents because they all live in a studio or a one-bedroom apartment. Parents of these students are mostly laborers who do not have time for them. Some students have to earn extra money by working at night. The public high school under study has had to deal with teenage pregnancy almost every year. Hardcore punishments such as expunging pregnant girls or drug users from school are not the best solution according to the sex education teachers at this public high school. These youngsters will create more problems for the society in the long run. The solution of the private high school to allow a pregnant girl to come back to school after giving birth is much better. This author interviewed a girl in the private school who was allowed to come back to class after she had given birth to a child. She was very grateful to the teachers. She realized that her boyfriend, with whom she eloped, had a low level of education and could not make ends meet. Living surrounded by drug dealers and users made the girl reflect upon her future and that of her child. She decided to get help from her parents and the school and fortunately was welcomed back by the school. She said she wanted to pursue higher education and had learned from her mistakes.

According to students, do the mass media, such as TV or radio, have any effect or impact on their lives in anyway?

The majority of the students under study consider sensational news, especially news related to rape cases, normal. They agree that news reporters should use sensual worlds to describe the rape acts. They considered the branding of a rapist, such as a lustful teenager or horny teacher, appropriate. They explained that that would scare future rapists. For the victims, students agreed that their real names should be fictitious and their pictures should be pixilated. Some state college students commented upon the censoring of images of people smoking on TV by only pixilation of the cigarettes as ridiculous. Most high school and college students like to watch a Thai soap series in which the female main character would slap the face of the main male character and the male would kiss her to get even. Some drama series portray the male character as raping the female first before they fall in love. Students explain that these kinds of drama and soap are available on every TV channel. The students under study stated that they could differentiate the soap from real life situations. However, they admit that young children also watch them without parental guidance. Mostly female high school and college students enjoy this kind of theme. They did not realize that it can instill domestic violence. Most students felt indifferent to how the most circulated newspapers post photos of sexy women in bikinis with R-rated comments.

Most students watch Hollywood movies and consider that the Western lifestyle portrayed in the movies influence their sexual behaviors. Some students admitted that they watched pornography via websites or from pirated DVDs. Students complained about Western tourists who touch, hug and kiss each other in public as setting bad examples for Thai youngsters.

How did the new media, such as the Internet or the social networking programs, influence the dating pattern and relationships of Thai students?

Almost all students under study are Facebook and Hi-Five users. Some students met new friends and use the social networking program to make appointments during the weekend. Some students have been drinking and going to pubs since they were 15–16 years old. They said there was no regulation for underage students. Students use cell phones. They would spend unlimited hours in front of the computer at home. Parental control was hardly ever heard of. Some private high school students stated that they missed out sending their homework on time because they were Facebook addicts. Some students at the public high school under study admitted to being online gaming addicts. One student said that he did not go to his former school at all. He went to a game shop and stayed there the whole day without telling his parents. A few students in this school were expelled because they posted a clip of a couple having sex in the toilet of the school on YouTube. Some students use the program Camfrog to chat about sex online.

Conclusion

From this author's investigation, there is a lack of understanding about the socio-cultural aspects that impinge on individual risk-taking behavior among students. Findings reveal pitfalls in Thai sex education and education as a whole. Lack of self and culture norm, critical reflexivity, basic knowledge of STDs, basic understanding of the social, parental cooperation, adequate living conditions, proper sexual consultation/guidance, compassionate and liberal school policy, effective curriculum, qualified and dedicated sex education teachers and enough media for sex education are putting Thai youths at risk of teenage pregnancy, HIV, STDs and drug use. A student-centered education is not yet happening due to the hierarchical structure of the Thai culture.

All of the above conclusions indicate that Thailand has not achieved sustainability in light of social injustice and power distribution as indicated by the HIV/AIDS issue discussed in this chapter and assessed by Jabareen's ten concepts of a theoretical sustainability framework.

Therefore, it remains to be seen whether the Thai socio-cultural context regarding sexuality and gender relationships is sustainable and equitable in the long term.

References

D'Agnes, T. (2001). *From Condoms to Cabbages*. Bangkok: Post Books.

Elliot, A. & Lemert, C. (2006). *The New Individualism: The Emotional Costs of Globalization*. London and New York: Routledge.

Friedman, T. L. (2007). *The World is Flat*. Picador/ Farrar, New York: Straus and Giroux.

Galarraga, O., Colchero, M. A., Wamai, R. G., & Bertozzi, S. M. (2009). HIV Prevention Cost-Effectiveness: A Systematic Review. *BMC Public Health 2009*, 9 (Supp I): SS doi:IO.I I86/147-2458-9-SI-SS, 1–14. Retrieved 13th June, 2010, from http://www.biomedcentral.com/I47-2458-9-SI-SS

Hassan, R. (2008). *The Information Society. Digital Media and Society Series*. Cambridge: Polity Press.

Hollander, J. A., Renfrow, D. G., & Howard, J. A. (2011). *Gendered Situations, Gendered Selves*. Rowman & Littlefield Publishers, Inc: N.Y.

Hopkins, A. G. (2002). Introduction: Globalization – An Agenda for Historians. In A. G. Hopkins (Ed.), *Globalization in World History* (pp. 1–10). London: Pimlico.

Jabareen, Y. (2011). Teaching Sustainability: A Multidisciplinary Approach. *Creative Education*, 2(2), 388–392.

Klausner, W. (1997). *Thai Culture in Transition*. Bangkok: The Siam Society.

National AIDS Prevention and Alleviation Committee (Jan 2006–Dec 2007) *UNGASS Country Progress Report-Thailand Nonthaburi*, Thailand: Department of Disease Control, Ministry of Public Health.

Malikhao, P. (2007). *HIV/AIDS Prevention Strategies in Two Thai Communities: Buddhist and Christian*. A PhD dissertation submitted to The University of Queensland http://espace.library.uq.edu.au/eserv/UQ:151849/n40871457_phd_front.pdf

Malikhao, P. (2012). *Sex in the Village: Culture, Religion and HIV/AIDS in Thailand.* Chiangmai & Penang: Silkworm Books and Southbound Publishers.
McQuail, D. (2005). *McQuail's Mass Communication Theory.* Fifth edition. London: SAGE Publications Ltd.
Nash, K. (2000). *Contemporary Political Sociology.* London: Blackwell Publishers.
PATH (2007). *Handbook of Sex Education Organizing Process for Youth for 7–9 Graders.* Bangkok: PATH.
Scholte, J. A. (2005). *Globalization: A Critical Introduction.* Basingstoke: Palgrave Macmillan.
Stifel, L. D. (1976). Technocrats and Modernisation in Thailand. *Asian Survey,* 16(12), 1184–1196.
Taylor, P. (1997). *Investigating Culture and Identity.* London: HarperCollins Publishers Ltd.
Thompson, J. (1995). *The Media and Modernity. A Social Theory of the Media.* Cambridge: Polity Press.
Tomlinson, J. (2001). Cultural Globalization and Cultural Imperialism. In F. J. Schuurman (Ed.), *Globalization and Development Studies: Challenges for the 21st Century.* London, Thousand. Oaks and New Delhi: SAGE Publications.
Twenge, J. M. & Campbell, W. K. (2009). *Living in the Age of Entitlement. The Narcissism Epidemic.* New York: Free Press.
UNGASS (2008). Country Progress Report Thailand. Nonthaburi: Office of Technical Development to Support HIV/AIDS Responses, Department of Disease Control, Ministry of Public Health http://search.unaids.org/search.asp?lg=en&search=ungass%20country%20reports, accessed on 1st March 2012.
UNDP (2004). *Thailand's Response to HIV/AIDS: Progress and Challenges.* Bangkok: United Nations Development Program.

Websites

Center of Disease Control, Ministry of Public Health, Thailand: http://aidsthai.org
PATH: http://www.path.org
Daily News (Thai newspaper) online: http://www.dailynews.co.th 19th December 2006, accessed on 5th February 2007.
Manager Online: http://www.manager.co.th, 10th February 2007, accessed on 12th February 2007; http://www.manager.co.th, 10th September 2006, accessed on 11th September 2006.
Teenpath (Sex Education online for Thai Youth): http://www.teenpath.net; http://teenpath.net/teenpath/database/Coverage-Province.asp, accessed 22nd October 2009; http://teenpath.net/teenpath/ database/Summary.html, accessed 22nd October 2009.

9

Factors that Stand in the Way of Green Communication in Africa

Henri-Count Evans

Introduction

The green communication discourse is housed in the domain of communication for sustainable development. Sustainable development is seen as a process of change guided by a number of values or principles. A strategic document entitled "Caring for the Earth", a strategic plan for a sustainable future prepared by the International Union for the Conservation of Nature (IUCN), the United Nations Environment Programme (UNEP) and the World Wildlife Fund (WWF) (1989, 2000) defines it as: "A kind of development that provides real improvements in the quality of human life and at the same time conserves the vitality and diversity of the Earth. Living sustainably depends on a duty to seek harmony with other people and with nature."

The guiding rules are that people must share with each other and care for the Earth. Humanity must take no more from nature than nature can replenish; this in turn means adopting lifestyles and development paths that respect and work within nature's limits. It can be done without rejecting the many benefits that modern technology has brought, provided that technology also works within those limits (IUCN, UNEP and WWF 1991).

Development policies, expert and academic presentations focusing on development in general, and climate change in particular, emphasize the need for increased awareness in respect to climate change problems. This emphasis on awareness makes relevant, and justifies the need for effective green communication and hence the enhanced role of the media in mainstreaming climate change discourse. Green communication is the vehicle via which information on the presence, impacts, adaptation and mitigation of climate change can be disseminated efficiently. Green communication is a new media genre that is directed toward climate change awareness, with the main agenda being to conscientize communities on the presence of climate change and educate them on how they can adapt and mitigate the negative impacts of climate change.

This in a broader sense helps to sustain the present and future generations against the traumatizing dangers promised by the changes in climate. Green communication will give people more power to control nature and lead in the transition from controlled change to changed control. Shifting the role of the media from its orthodox role in society involves a lot of work from scholars and policy makers. There is need for a complete paradigm shift from the way climate change issues are reported and interpreted in the media. That is, if they are given the green light to be reported about. It is even more difficult in Africa to mainstream the green communication genre itself among media practitioners before moving to mainstreaming climate change. The African continent in its totality is the least prepared to embark on an environmental evangelism and preach the climate change discourse, but at the same time the worst affected by the climatic hazards.

Dawn has come for Africa to identify the critical role of information, not only in terms of the print media, but also other communication media such as broadcasting and the Internet, and also the emerging usefulness of narrowcasting in shaping development agendas in the world, for example, the use of the telephone networks to send development messages to diversified subscribers. These mediums must be made accessible to the people who are at the heart of missing the climate change story.

This chapter's point of departure is that the media are central to climate change awareness as primary sources of public information and also as agents of representation and social change. The news media function specifically as an authoritative version of reality that specializes in orchestrating everyday consciousness for the public (Gitlin 1980). Green communication as discourse practice requires an informed citizenry that is capable of interpreting climatic and weather patterns that will enable them to take informed decisions with regards to their environmental actions as individuals, groups and greater communities.

The media is regarded as the most effective leeway to regularizing the green revolution in society. The accessibility, immediacy and intrusiveness associated with the media are critical to the effective mainstreaming of the climate change story. A careful appreciation of the media's role, in mainstreaming climate change in Africa, need to take place within a critical understanding of the contemporary African communicative contexts.

Already confronted with bad climatic impacts, Africa needs to radically restructure its communicative terrains from the traditional politically biased reportage to a policy of frontal encouragement that targets benefiting green communication. Against the Broadway of hazards posed by climate change, Africa needs to declare war on climate change, a confrontation whose contours will definitely change the continent's environmental landscape. This chapter refers to the African context, not in the sense that all African countries share the same internal structural dynamics, the term *Africa context* is used for a "merely geographical" convenience to start a point of analysis, in the same logic that Annabelle Sreberny (2008) used the term "Middle East" for a linguistic convenience to demarcate a point of analysis.

Background

The United Nations Framework Convention on Climate Change (UNFCCC 2000) defined climate change as a change in climate that is attributable directly or indirectly to human activity that alters atmospheric composition. However, it can be noted that climate has always varied naturally, but the changes that have been noted in the last century have outpaced the natural variations, which occur over larger time scales (ibid).

Scholars and scientists within the climate change spectrum concur that the recent manifestations of global warming indicate something exterior and unnatural, something driving the climate change gear beyond the normal speed limit. The amount of greenhouse gasses in the atmosphere is beyond the capacity of the available sinks. This has and is worsening the ozone depletion, thereby exposing the earth to ultra-violet heat waves that are anti-plant, anti-animal and anti-human existence.

Climate change will have disastrous outcomes especially for Africa. The change will result in the transformation of the current patterns of agriculture, relegating some districts to semi-arid regions and semi-deserts. This is likely to force population pressures into the productive areas, causing socio-economic strife and even civil and trans-border conflicts. This will also increase human poverty and starvation. It is anticipated that climate change will cause water stress. Crop and animal production are also projected to suffer from the changing climate impacts as areas suitable for agriculture will diminish.

The length of the growing season and yield potential, especially along the margins along the semi-arid and arid areas are expected to decrease. This exposes the lives of people and humans to disaster due to recurrent droughts. Droughts will adversely affect the continent's food security and exacerbate malnutrition.

Communication infrastructure

Any communicative action requires that the communicator reaches a certain audience in order to make the whole communicative process a success. Reaching the target audience is predetermined by the availability of the means of communication that best suit the essence of that communication. That is, the requisite communication pathways or technology.

This may include broadcasting signals, satellite signal transmission infrastructure, cable transmitters and fiber optic networks. Africa's communication infrastructure is a concern in achieving positive climate change mainstreaming goals. As a Third-world continent, Africa is still lagging behind techno-wise. Communication networks are located in urban areas where only about 20 per cent of the total population live.

Broadcasting

In terms of broadcasting, the television signals are only accessible within certain radii, marginalizing most rural areas. This is notwithstanding the fact that broadcasting has proved in other parts of the world to be useful as a channel for mainstreaming climate change. In Africa, on average, the television signals are only accessible to 56 per cent of the population. This has seen the further widening of the digital divide. In such circumstances, it is only logical to dream of green communication. Why? Because the broadcasting infrastructure necessary to facilitate and convey green messages is hardly available to larger populations. Evans (2010) notes that, "communication without proper means of communication is no communication at all".

For example, Zimbabwe's south and western provinces do not have television and radio signals. This leaves citizens with no alternative sources of information, and hence symbolically annihilates them and renders them foreigners in their own country, as they cannot participate in any meaningful national debates and development processes. Such citizens are left out in a complete media and information blackout. This scenario makes any effort toward communication for social sustainability a non-entity. Any efforts toward communication for sustainable development should first take care of this problem to make the communication process realistic and achievable, besides that there will be no communication to talk about.

Moreover, where signals are accessible, very few people have surplus money to spend on buying a television or radio set, which in a typical African context is regarded as luxurious. The rural populations are more occupied with issues to do with hunger and poverty alleviation and hence put less priority on leisure and entertainment products. This further complicates green communication initiatives. At the end of the day, very few people have access to the media, which means even if the information on climate change is made available it reaches very few people and makes no major impact.

The lack of broadcasting infrastructure is a major barrier to productive communication in Africa. Almost 80 per cent of the population live in rural areas. In such areas, electricity and telephone lines, and fiber optic networks are rarely known or heard of. The issue of electricity unavailability makes it impossible for even those with the money to purchase television or radio sets. The only communities with electricity and telephone lines are those at growth points and business centers who constitute a meager percentage of the rural populations.

It must also be noted that most African nations have access to satellite signal transmission especially through DSTV-Multichoice. This is a welcome platform where climate change debate can be deliberated and help create a responsible citizenry. But again, there are prohibitive factors regarding accessibility. Firstly, the cost of installation is beyond the reach of many. As if that were not enough, the monthly subscription charges are out of harmony with logic when considering the fragile African economic context. As such, few people in the towns can afford to purchase and install the satellite dishes. To further complicate the situation, the most synonymous language used throughout the satellite channels is predominantly English and French. Given that the literacy rate stands at 56 per cent in Africa, it makes sense to conclude that only about half the population can grasp the language barriers.

New Information and Communication Technologies (NICTs)

Furthermore, the issue of an information society and information superhighway is still far from a reality in Africa. The problem starts with the lack of policies on information technologies on the part of governments. The term "*information society*" is hardly applicable in any African context. Even though there is a lot of information about climate change on the Internet, very few of the elite in urban cities and towns have access to it. To worsen the situation further, very few people even in the towns are computer literate. Some scholars posit that the computer literacy rate cannot be more than 3 per cent. Estimates indicate that Africa is technologically behind the developed countries by more than 23 years. If quantified, a gap of this magnitude makes much of the technological discourse in Africa irrelevant.

It is within this context that Africa needs to discover the urgency of mainstreaming climate change looking at the hazards which are being determined by the changing climate. It will be fatal for the continent to overlook the urgency of the moment and engage in the luxury of cooling off. As the examples above have indicated, green communication is far from being actualized in Africa, regardless of the continent's thirst for information.

The emergence of the Internet and other information and communication technologies gave rise to media convergence; electronic systems have been connected to text, sound and pictures. Communications have been made faster, firmer and deeper. Friendships, networks and associations are being established and experienced over the Internet. Many people can establish relationships that were formerly private. As information and communication technologies have connected audio, video and electronic texts into one network, NICTs reduced people's dependency on mainstream media such as traditional newspapers, television and radio. The Internet in essence can be said to have introduced an alternative public sphere where groups and individuals in the "global village" can share information on issues that impact their daily lives, such as climate and environmental information.

As such, climate change topics have been introduced online through the creation of climate change websites, networks and blogs. The subject is also topical on social associations such as Facebook and Twitter. This shift from mainstream to alternative media platforms has serious implications in regard to climate change mainstreaming in developing Africa. The concept of the "Global Village" is yet to gain technological relevance in the African context.

The technological proliferation has widened the digital divide between the North and South. Noting that the current global mode of production is greatly informed by "knowledge", the society is now "knowledge-driven", "information-driven", and "discourse driven"; all these developments taking shape on the penetrative and interactive Internet. Africa in particular and the developing countries in general have been left out of social development. Scholars note that scientific ignorance is an embedded and expectable feature in the knowledge society. While the idea of the "information society" has become a master metaphor of our time (Martin 1995), there is an underside of this transformation in Africa that requires examination.

That there has been rapid proliferation of scientific and technical knowledge, as well as other forms of data and information, is incontestable. But this information explosion and the associated revolutions in technology imply and in fact necessitate the social distribution of ignorance. Lukasiewicz (1994: 122) observes that with the exponential growth in the volume and complexity of information one's degree of grasp (the ratio of information the human intellect can handle to the volume of information available) "is quickly diminishing while one's (degree) of ignorance is on a fast rise".

Even though the proponents of the "information society" theory posit that the NICTs liberated communications making them available and accessible to all for much less, this benefit is yet to be realized and extended to the developing South. With the rapid resurgence of citizen journalism claiming its toll in the information age, it was hoped that information would be made available faster and without accessibility constraints with regards to laws and regulations, but it is still more of an abstract idea rather than a fact in Africa.

It must also be noted that the advent of the information technologies has implications for green communication in Africa. The inadequacy of basic telecommunications infrastructure and electricity reliability is hindering the rapid expansion of new information technologies, especially computer technology, in the Third World and Africa in particular (Melkote 2008). The individual access to interactive technologies is unevenly distributed, where there are discrepancies between the North and the South, the rural and the urban, the elite and the working class, and the rich and the poor. Because of the centralized nature of communications infrastructure in Africa, communication technologies tend to be enjoyed by a few urban elites who have the money to buy a computer, for example. As such most people are left out of the information society.

There is an argument that communication technologies have widened the information gap between those who were or are already socially privileged and economically sound, and those who were or are not. This imbalance is scalar in the sense that it is at global, regional, national and at local levels. McQuail (1994: 90) argues that this has created a new "information underclass" rather than an information society. Sussman and Lent (1991) argue that the new technologies have redefined the centers to mean transnational corporations that have made affected nations more dependent on them not only for technology, but for other economic needs as well.

The politics of green communication

The media in general are part of the ideological tools that are used by nations and political blocks to legitimate their hegemony in society. Some scholars argue that since their inception media have been used as political tools meant to serve the political interests of those in power.

Green communication stands vulnerable to political definitions. It is unfortunate that the concept of green communication is being introduced in the midst of a complete ideological dilemma. The climate change discourse has seemingly been a preoccupation of the global media. Very little has been done by the national media and nation-states at their own scalar levels to address the climate change problem. As such the global media have been perceived through a negative lens by the developing Africa due to their historically entrenched news flow imbalance. The images transmitted by these media are typically married to the stereotyped portrayals that developing countries have always been subjected to.

Opinion leaders in Africa have developed a negative perception with regards to the way they interpret and internalize international news. Green communication, while operating at a micro-level of the ideological minefield, has been affected (Evans 2010). The green texts from the international news media tend to signal ideological standpoints of the originator and disqualify oppositional beliefs. This chapter notes that green communication has emerged as an alternative global sphere where the supra-impositions of ideological superiority have acquired a sharp presence.

This has seen the re-introduction of *cold war news management techniques*. There is now a green global clash between the architects of the current global order vis-à-vis the anti-system protagonists. The relationship between transnational media and their multinational industrial corporations in Africa is also a cause for concern. The media turn a blind eye to the environmental damage caused by these giant corporations as this would compromise their operations in Africa. Instead, often the media have portrayed the activities of these corporations as eco-friendly. This has effectively worked in regularizing their stay in business. In some instances these companies have used clandestine ways of winning public support by hiring unorthodox scientists to create confusion with regards to origins and impacts of climate change. These pseudo-scientists have found space in the global media where they come out disputing the human factor in the climate change process. This has happened against widespread scientific consensus on the urgency and impact of the changing climate.

The green communication discourse has been mired in this ideological manipulation. The presence of multinational companies and their sister international media corporations and news agencies have impeded on the successful mainstreaming of climate change in Africa and the developing countries in general. Companies and politicians with vested interests in maintaining the status quo are in the process of undermining science and subverting journalism (Fairclough 2003). By doing so, the media have maintained the multinational companies' control of the economic structures in Africa.

Economic development and environmental protection

Before concluding this chapter it is also useful to note that the rhetoric of enlightenment and industrialization have also made climate change communication a difficult task. The developing countries are more concerned with development of their communities and much of this development is more centered on activities that are anti-environment. The proliferation of industries in Africa complicates the whole situation. More greenhouse gas emissions are being emitted by these industries that are huge manufacturing, agriculture and mining corporations. This has been further compounded by the use of outdated manufacturing and farming methods that worsen environmental destruction.

The proliferation of industries also meant the clearing of more forests to accommodate these developments, which in the long run destroy the ecosystems. This is slowly but surely leading to desertification, with some regions in the Sahel having become victims already. So the drive to development should also be seen from the same lens as a factor impeding the green communication initiatives, as people are preoccupied with economic progress at the expense of environmental concerns.

Some industrial corporations' images are compromised because of the effects they have on the environment when exploiting natural resources, energy sources and manufacturing. Most of these companies have become pillars of the environment. This make-up is often backed by the so-called Corporate Social Responsibility (CSR) principles. Through advertising, sponsorship and public approaches, they make people believe that major companies are environment-friendly and that they even promote world sustainability. This has also been exacerbated by the lack of institutionalized campaigns by local and national authorities, whose duty it is to generate civic awareness. They left environmental defense in the hands of industrial corporations.

Political economy of the media in Africa

The operations of the media are always over-determined by their relationship with the existing mode of production and their ownership and control structures. This relationship between the system, owners and the media practitioners influences the texts to be produced and hence direct society toward the preferred angles of the owners. This is also applicable when analyzing the impact of ownership and the green revolution. A valid analysis of factors that stand in the way of green communication in Africa will not be complete without a critical examination of ownership and control patterns and also media funding sources.

There are basically two kinds of media in Africa: public and private.

Public/state-owned media

Public media institutions are those media institutions that are publicly owned and funded and are there to serve the interests of the public. However, this notion of public media in Africa remains idealistic from an African point of view. The so-called "public-media" are in actual fact "state media" that basically serve the interests of the government of the day. These media are generally funded by the state and are usually put under the control of the ministry or department of communications and information. The state and not the public have more power over the editorial policies of these media houses. The board of directors are normally appointed by the minister responsible for communications. In this instance, the media are likely to adopt news management strategies that favor the state and legitimize government policies and points of view.

Most post-independent African governments inherited this system from their colonial predecessors. Since its inception, the media in Africa has been a contested terrain with those in power seeking total dominance and control over the discursive platform. Despite claims to neutrality by and the rhetoric of Sierbert's media "for development" normative model, the ruling elite has always used the media as a tool for political manipulation and an agent of social change, as conceptualized by Adorno (1991) and McChesney (2000).

African governments discovered the enormous potential of the media in centralizing their hegemony. They have, through the rhetoric of public media, sought to neutralize and pacify counter-hegemonic forces in society by bottlenecking the available media space. These strategies have worked and made citizens subjective to a view that accepted their rule as legitimate and hence extended their stay in power. Through draconian media laws and regulatory mechanisms, those in power have controlled what people can read, listen to or view, thereby effectively controlling public consciousness.

With particular reference to Zimbabwe, the state enjoys a monopoly over broadcasting through the state parastatal Zimbabwe Broadcasting Holdings, which owns the Zimbabwe Broadcasting Corporation television and four radio stations. This kind of monopoly has serious green communication implications. The public scope of environmental reference is grossly limited. The state-run institutions are therefore not in a position to ridicule and criticize the policies of government with regards to environmental decisions as this will compromise their relationship with the owner. The public sphere is limited with the state media being more preoccupied with politics and business thus leading to "gutter journalism". Public environmental campaigns find little room in the news casts as they are relegated to second place after politics and economy.

Private media

The concept of private media posits that the private media are those media institutions that are privately owned and are commercial in nature. These media are also believed to be the ones that should act more as the public watchdog, guide dog and guard dog. Due to their independence from the government, they are perceived to have more liberty to independently report on issues of public interest. As with the private media, the situation is more or less the same with their state-run counterparts. The media space in most African countries is highly polarized with the state on the one hand and the opposition and the pro-opposition non-governmental organizations and the civic society organizations.

Most private media institutions are owned and funded by Western donors and some global media institutions. These ownership structures created and established news factories which brought under their control the power to decide what the public of each nation would be allowed to know of other nations and in what shade of meaning. This has seen the private media in Africa ignoring scientific consensus that indeed the subject of climate is not a product of the developing countries but rather is a phenomenon that traces its roots to the industrialized nations. Climate change and the destabilization of the environment are problems that do not originate in rural areas. Neither do they originate in less developed world areas.

The private media have danced to the tune of the multinational companies and seem to preach the environmental evangelism preached by the global media that in the final analysis serves to save the existence of neo-imperialism and capitalism. The private media in Africa are yet to appreciate that the main cause of climate change is capitalism. They are yet to portray the capitalist historical and development contradictions of development and environmental protection. They have rather paid their appreciation to the lip service of American and Western promises to be part of the mitigation and adaptation team. Through selective application of news values they have not reported on the Western double standards with regards to climate change issues. The refusal by the United States of America to sign the Kyoto Protocol did not find much coverage in the private media sphere.

The construction and selection of news strategies deployed by the private media is intrinsic with their monetary marriage with the fathers of capitalism. This has greatly impacted the development of green communication, a fundamental unit in mainstreaming the green revolution and effective communication for sustainable development as per the modification by Servaes (2008). In this scenario, it is appropriate to invoke the critical political economy of the media as a useful paradigm in demystifying media green representations. The notion of "who pays the piper dictates the tune" is relevant in analyzing how control and ownership patterns of the media are serious impediments to the successful realization of green communication targets and eco-friendly communications.

The private media have claimed to be agents of representation who represent vulnerable members of society from repressive institutions. However, they have symbolically failed to represent the African communities at the global arena by exposing the repressive eco-imperialism by the developed West. Their watchdog role has not been extended to cover the complex, unsustainable and subjective relationship between Africa and the developed world.

Media, commercialization, profits and green communication

With increased commercialization of the media, the inclusion of green journalism in the media is based on whether it is seen as commercially viable or not. Such considerations limit the green communication discourse, as well as full and authentic expression. For content to be commercially viable, it is limited to a few tried and tested genres of politics, business, entertainment and sport that can be sold to bigger markets. In a sense, therefore, the more removed from profitability green communication in modern day public communication, is the less likely it is to be selected for use in the development of media content (Martin 1995).

The media have shunned their traditional roles of interpretation, representation and fourth estate in society in favor of more trivialized news flavors that tend to be more sensational and entertaining. This has been made rampant by competition for audiences. This also is a product of a commoditization suspicious relationship between the media and business. Most media organizations have acquired a commercial character that subjectivized them to the economic tenets of business, especially the need for profit maximization. News contemporarily is now a commodity to which a commercial value is attached. This development re-directed the media to cover the best-selling stories that will help keep them in the news marketplace.

This also gave birth to filthy journalism or "dissident journalism", such as tabloid journalism that only focuses on the sexual, the sensational and scandalous, and in some instances sports. Even the traditional media have been forced into this route in order to remain economically relevant and viable. The tabloidization of the media is another contributing factor toward the cooling off of the eco-cultural façade. As a result there is public space deterioration (Glynn 2000), a lack of critical judgment (Boykoff 2004), and maybe, the environmental crisis (Kellner 2005; McChesney 2000).

Entertainment and leisure are more mighty matters in the media, while more serious information is filtered through expressions and formats that make the media lose value and impact in articulating issues that are in the public interest. More serious issues such as the changing climate are reduced to entertainment or edutainment which has become pervasive in the African and the general media landscape. The media content tends to be about, and speak to, the concerns of the individual rather than of society (Diaz Nosty 2007). Gitlin (2005) observes that the public exposed to the audiovisual "happy culture" is more likely to be manipulated. Their critical judgments are less developed and they show less signs of reflection over what they see in their public space.

The stimulating awareness is greatly diminished by sensationalist news media contents and poorly outlined media practices (Diaz Nosty 2007). Mainstream media now employ a pattern of reality presentation similar to that what in an earlier era would have been labeled sensationalism. The media do not address the climate change problems through relevant sources of information. They normally find a catchy headline. They often seek to create debate between two entities (ibid).

Far more from scientific evidence the media cover far-fetched perspectives, which are not part of the scientific consensus on the impact of climate change and can give away to skepticism and denial of the climate change problem (Weaver 2003). This makes the media not really an appropriate public arena on which to articulate the climate change problem, as they are more intimate with their economic incentive in selecting profitable news from the non-profitable news.

The more powerful industrial giants keep an intimate intercourse relationship with the media through their publicity investments that also keep the media in business. The media sensationalize eco-news through invoking their traditional news values of conflict "when it bleeds it leads"; conflict sells more than consensus. In the final analysis the excessive commercial usage of the media invalidates responsible agenda points and the right of information. Sensationalism and media cooling off maintains a vague relationship with green communication. The opinions it feeds are indifferent, uninformed or not committed to sustainability. This lack accentuates the relevance of new communication policies and the role they must play in adjusting media toward sustainability.

Media personnel training and climate change

Most reporters in Africa are not yet conversant with the climate change discourse and have not yet appreciated the green communication paradigm as a viable news platform. This problem emanates from the inadequate training offered at journalism schools. A pilot survey of training curricula will expose that there is not yet progress with regards to specialized reporting, especially on scientific journalism and climate change/green communication in particular.

Lecturers at most institutions are yet to grasp the need to mainstream green communication into the dominant training and teaching paradigm. This negatively impacts on the capacity of the media to report accurately on environmental problems. Reporters in most newsrooms are ill-equipped to tackle the complex and seemingly scientific subject of climate change. Green communication assimilation into the orthodox reporting styles will take time as both the training institutions and media institutions have a maladaptive capacity to internalize climate change as a serious news topic that requires both academic and practical attention.

For example, in Zimbabwe, out of the three major journalism training institutions, – National University of Science and Technology, Harare Polytechnic College and the Midlands State University – not one of them offers green communication or environmental communication as a specialization. The situation is the same in the media institutions themselves, all the mainstream newspapers, both private and state owned, are yet to introduce the "environment desk". A critical examination of these phenomena reveals that the environment section is not seen as one of the important sections. Furthermore, the media usually find it expensive to run such a section as it does not bring any income. If ever there are capital constraints these environment correspondents are the first to be eliminated.

Another curtain-raiser topic is that the non-governmental organizations that claim to be in the business of mainstreaming climate change are yet to convene workshops for academics and reporters where climate change training is undertaken. This leaves a lot to be desired with particular concern to mainstreaming climate change in Africa. In instances where ill-trained reporters attempt to raise public awareness on the green revolution, the information is generally not accurate from a scientific viewpoint. Also, reporters in this field have carried with them the journalistic ethic of balance. This has seen them trying to balance scientific consensus with skepticism. In most cases the issue is taken to the streets looking for pros and cons. This has made the questioning of scientific journalism in Africa more relevant. There is poor specialization in this field. Media events are not related to scientific events but to political events (Boykoff & Roberts 2007). Lack of scientific specialization prevents reporters transmitting scientific information rigorously (Russell 2008). The information in the media lacks consistency; it fluctuates according to news values in a way that climate change reports are deemed secondary after different news segments and current affairs (Roser-Renouf & Nisbit 2008).

Language decentralization

Another of the factors that stand in the way of green communication is the issue of language. This chapter categorizes language into two domains. One that talks of language from the literal meaning of the word in English, French, Arabic and so on. Secondly, language as the scientific jargon that is used by scientists in explaining climate change issues in particular and other scientific information in general.

The climate change discourse has been articulated in Africa through the so-called business languages, which are in actual fact colonial languages. The discourse has been communicated predominantly in the English language paradigm for English speaking African communities and in French in former French colonies.

This has and is still happening, notwithstanding the deplorable literacy rates in Africa. Most people, especially those in rural settlements, which constitute 80 per cent of the total population are highly illiterate and can neither speak, read nor understand these languages. They are more familiar with their local or vernacular languages; a fact that calls for the re-direction and decentralization of languages used in green communication to make the discourse applicable and relevant to African contexts.

The use of English in communication marginalizes those communities that speak a different language. Using local languages is also one of the essentials of communication for sustainable development (Evans 2010). The participation of these communities is necessary in fostering their adaptive and mitigatory capacities, which in essence needs a thorough understanding of the subject.

Information on the Internet and in the mainstream media is predominantly in English, which makes it useless for many. This has negative implications on the internalization of media texts by such societies. As communication is a two-way process between a sender and receiver, both need to understand the language used in the communication process in order to make meaning of the whole communicative process. The scientific language that scientists use to explain the climatic transformations is also a cause for concern for most people in Africa. Usually the information on climate change is encoded in scientific jargon not popular to ordinary interpretation, hence making the decoding process a mountainous task for the audiences.

Scientifically jargonizing the climate change discourse is discriminative as it excludes certain members of society from the climate change discussions. The language used demarcates between the prospective readers with those scientifically challenged being relegated as non-actors in a discourse that calls for inclusion and not exclusion. The power that language plays in society is immense as it can appropriate certain information to certain groups and not others.

This further widens the gap between the haves and have-nots, rich and poor, elite and middle class, north and south; a development that is not healthy for a fruitful scientific discussion. The entrenchment of the climate change discourse in scientific terrains makes it difficult for the media to decode the messages accurately and can lead to public misinformation. The art of entrenching the climate change discourse in the scientific language makes communication monologic, anti-dialogic and anti-discursive. The scientists must, therefore, keep it simple (Evans 2010).

Research on the knowledge gap paradigm reveals that prior knowledge in an area is critical to assimilate new information in that area (Viswanath & Finnegan 1996). Starting with conceptual anchors, the gaining of knowledge in a field tends to follow a spiral model, with new bits added to prior accumulations. But as proliferation conceptual anchors and technical terms are overlaid with new facts and modifications, knowledge domains become forbidding to outsiders.

All but the most persistent non-specialists are effectively restricted from the discourse discussions and lobbying (Royce 1997), thereby creating the "us" versus "them" relationships in society. Scientific ideas are encoded in a distinct language and need to be decoded to be accessible to the public. Thus high entry qualifications into the climate change discourse militate against the internalization and acquisition of scientific knowledge. This also makes claims to a discourse, knowledge and information driven society grossly inapplicable when people cannot access the information in the discursive structures needed to gain subject knowledge.

Conclusion

This chapter has noted with concern the terrible profile of climate change issues in African media, both in the public and private, electronic and print media. The media in Africa exhibits a serious case of ignorance when it comes to scientific journalism. The chapter traced the problem back to the core journalism mentality factories (training institutions), that is, the training institutions themselves. The journalism training departments have nothing to offer with regards to green communication.

There is little scientific specialization in Africa. This has serious implications for the final output produced by the media. The training institutions are yet to take climate change as a serious journalism training paradigm. Most journalism students exhibit total climate change science ignorance and a complete lack of interest in communication for sustainable social innovation.

In the news media institutions, most journalists and media managers lack interest in the climate change discourse. In terms of deciding what is newsworthy, economic determinants of profit and maximum sales are used as the news selection indices. Climate issues, because of their sheer lack of drama, explicit urgency and obtrusivity, are sidelined in favor of the more common genres of politics, sports and business.

The establishment of climate news desks in the newsrooms is considered an economic or financial risk. In terms of who writes climate change and environmental issues, anyone can do it. There are no recruiting policies that specifically require someone with prior qualifications and knowledge on climate change to be hired. This has resulted in wrong priming, angling and framing of climate change stories. Whenever climate change is reported about, the reports are notoriously inaccurate and fixed along the ideological matrix of a particular news organization. Climate change is only spoken about in the media with reference to a major environmental disaster or great speech by politician or interest group representative.

Audiences lack climate change awareness. Firstly, the media is not communicating climate change issues, hence the lack of knowledge among audiences. Secondly, if the media do communicate it, either the information is chronically inaccurate, or it is about a disaster that is happening somewhere else.

The change is seen as happening elsewhere and with nothing to do with "us". There is configuration within the minds of audiences that climate change is "their" problem and not "our" problem. This chapter has noted that there is a symbolic detachment of audiences' perception of climate change along the lines of the "them" versus "us" kind of scenario.

The other issue in connection to audience assimilation of climate change knowledge among the audiences is the issue of language. The mainstream media institutions in Africa use English, French, Portuguese and Spanish as the predominant communication languages. Some members of society, even though they can be said to have an elementary education qualification, are not conversant enough with these languages. Also, climate change being a scientifically entrenched paradigm is often communicated through the use of specialized science language. Those audiences with no climate change scientific language command are effectively left out of the green communication discourse.

References

Adorno, T. W. (1991). *The Culture Industry.* London: Routledge.

Antilla, L. (2005). Climate of scepticism: US newspaper coverage of the science of climate change. *Global Environmental Change,* 15: 338–352.

Barker, M. J. (2007). Global greens and the mass media: Building for a participatory future? *Griffith Journal of the Environment,* Issue 2 Article 2, 24.09.2007, http://www.griffith.edu.au/faculty/ens/gje/issues/2006-2/articles/gje-2006-02-barker.pdf

Baron, J. (2006). Thinking about global warming. *Climatic Change,* 77(1–2): 137–150.

Boykoff, M. (2004). Journalistic balance as global warming bias: Creating controversy where science finds consensus. *Fair,* November–December: http://www.fair.org/index.php?page=1978

Boykoff, M. & Roberts, J.T (2007). *Media Coverage of Climate Change: Current Trends, Strengths, Weaknesses.* United Nations Development Programme. Human Development Report 2007.

Diaz Nosty, B. (2005). *El déficit mediátic.* Barcelona: Bosch.

Diaz Nosty, B. (2007). el nuevo continente virtual. Diaz Nosty, B. (dir.), *medios de Comunicacion/tendencias. el escenario iberoamericano.* Barcelona: Ariel: 13–86.

Evans, H. (2010). *Climate Change in the Digital Age: Opportunities Lost and Found.* Bulawayo: Nust.

Fairclough, N. (2003). *Critical Discourse Analysis in Researching Language in the New Capitalism: Overdetermination, Transdisciplinarity and Textual Analysis.* Lancaster University.

FAO. (2007). *World Congress on Communication for Development.* Washington: World Bank Pub.

Gitlin, T. (1980). *The Whole World is Watching: Mass Media in the Making and Unmaking of the New Left.* Berkley: University of California Press.

Gitlin, T. (2005). *Enfermos de información. De cómo el torrente mediático está saturando nuestras vidas.* Barcelona: Paidos.

Glynn, K. (2000). *Tabloid Culture: Trash Taste, Popular Power, and Transformation of American Television,* Durham: Duke University Press, http://www.csicop.org/scienceandmedia/climate/

IPCC. (1998). *The Regional Impacts of Climate Change: An Assessment of Vulnerability.*

IPCC. (2001). *Climate Change 2001. Impacts, Adaptation and Vulnerability.* Chapters 10, 11, 17 and 18. Contribution of Working Group II to the 3rd Assessment Report of the IPCC.

Kellner, D. (2005). *Media Spectacle and the Crisis of Democracy: Terrorism, War, and Election Battles.* Boulder: Paradigm.

Ladle, R. et al. (2005). Scientists and the media: The struggle for legitimacy in climate change and conservation science. *Oxford: Interdisciplinary Science Reviews,* 30(3).

Lukasiewicz, J. (1994). *The Ignorance Explosion: Understanding Industrial Civilization.* Ottawa: Carleton University Press.

Martin, W. (1995). *The Global Information Society.* Ashlib: Hamshire.

McChesney, R. (2000). *Rich Media, Poor Democracy: Communication Politics in Dubious Times.* New York: New York Press.

McCright, A. M. & Dunlap, R.E. (2003). Defeating Kyoto: The conservative movement's impact on U.S. climate change policy. *Social Problems,* vol. 50, n. 3: 348–373.

McQuail, D. (1994). *Mass Communication Theory: An Introduction.* London: Sage.

Melkote, S. R. (2008). *Communication for Development and Social Change.* New Deli: Sage.

Nisbet, N. C. (2007). *A "two Step flow of Popularization" for Climate Change: recruiting opinion-Leaders for Science.* CSI [Committee for Skeptical inquiry], http://www.csicop.org/scienceandmedia/climate/

Nisbet, N. C. (2008). *Moving Beyond Gore's Message: A Look Back (and Ahead) at Climate Change Communications.* CSI [Committee for Skeptical inquiry], http://www.csicop.org/scienceandmedia/beyond-gores-message.

Roser-Renouf, C. & Nisbet, M. C. (2008). The measure of key behavioural science constructs in climate change research. *International Journal of Sustainability Communication,* 3: 37–95.

Royce, J. (1997). Surviving information overload: Lessons from the reading reluctance research. *School-Libraries-Worldwide,* 3: 39–46.

Russell, C. (2008). Climate change: Now what? A big beat grows more challenging and complex. *Columbia Journalism Review,* July–August.

Servaes, J. (2008). *Communication for Development and Social Change.* Los Angeles: Sage.

Servaes, J. & Carpentier, N. (2006). *Towards a Sustainable Information Society.* Bristol: Intellect Books.

Sreberny, A. (2008). "The Analytic Challenges of Studying the Middle East and its Evolving Media Environment". In *The Middle East Journal of Culture and Communication,* Vol. 11, Issue 4:8–23.

Sussman, G. & Lent, J. A. (1991). *Transnational Communications: Wiring the Third World.* Newbury Park: Sage.

Tabeaud, M. (2008). Concordance des temps. *EspacesTemps.net,* http://espacestemps.net/document4343.html. http://unfccc.int/essential_background/convention/background/items/2536.php.

Viswanath, K. & Finnegan, J. (1996). The knowledge gap hypothesis: Twenty-five Year Later. In Burleson, B (Ed.) *Communication Yearbook 19.* California: Sage.

Weaver, A. J. (2003). The science of climate change. *Geoscience Canada,* 30(3): 91–109.

Weaver, A. J. (2008). *Keeping our Cool: Canada in a Warming World.* Toronto: Viking.

Part IV

Conclusion

10
Future Challenges for Communication for Sustainable Development and Social Change

Jan Servaes

Communication for Sustainable Social Change and Development theory and practice have been changing over time in line with the evolution of development approaches and trends and the need for effective applications of communication methods and tools to new issues and priorities. Communication in sustainable development has addressed the specific concerns and issues of food security, rural development and livelihood, natural resource management and environment, HIV/AIDS prevention, poverty reduction and equity, and gender and information and communication technologies (ICTs).

Within this framework, communication is viewed as a social process that is not just confined to the media or to messages. Development Communication methods are appropriate in dealing with the complex issues of Sustainable Development in order to:

- improve access to knowledge and information to all sectors of society and especially to vulnerable and marginalized groups;
- foster effective management and coordination of development initiatives through bottom-up planning;
- address equity issues through networking and social platforms influencing policy making;
- encourage changes in behavior and lifestyles, promoting sustainable consumption patterns through sensitization and education of large audiences;
- promote the sustainable use of natural resources considering multiple interests and perspectives, and supporting collaborative management through consultation and negotiation;

- increase awareness and community mobilization related to social and environmental issues;
- ensure economic and employment opportunities through timely and adequate information; and
- solve multiple conflicts ensuring dialogue among different stakeholders in a society.

The sustainability of social change processes

In Servaes (2007, 2008) I subdivided communication strategies for development and social change at five levels:

- *Behavior change communication* (BCC) (mainly interpersonal communication).
- *Mass communication* (MC) (community media, mass media and ICTs).
- *Advocacy communication* (AC) (interpersonal and/or mass communication).
- *Participatory communication and social mobilization* (PC) (interpersonal communication and community media).
- *Communication for structural and sustainable social change* (CSSC) (interpersonal communication, participatory communication and mass communication).

Interpersonal communication and mass communication form the bulk of what is being studied in the mainstream discipline of communication science. Behavior change communication is mainly concerned with short-term individual changes in attitudes and behavior. It can be further subdivided in perspectives that explain individual behavior, interpersonal behavior, and community or societal behavior.

Behavioral change communication (BCC), mass communication (MC) and advocacy communication (AC), though useful in itself, will not being able to create sustainable development. *Participatory communication (PC) and CSSC are more concerned about long-term sustained change at different levels of society.*

Looking at *desired or expected outcomes*, one could think of four broad headings:

- approaches that attempt to *change attitudes* (through information dissemination, public relations and so on);
- *behavioral change* approaches (focusing on changes of individual behavior, interpersonal behavior and/or community and societal behavior);

- *advocacy* approaches (primarily targeted at policy-makers and decision-makers at all levels and sectors of society); and
- *communication for structural and sustainable change* approaches (which could be either top-down, horizontal or bottom-up).

The first three approaches, though useful by themselves, are in isolation not capable of creating sustainable development. Therefore, sustainable social change can only be achieved in combination with and incorporating aspects of the wider environment that influences (and constrains) structural and sustainable change. These aspects include (see also McKee et al., 2000): structural and conjunctural factors (e.g., history, migration, conflicts); policy and legislation; service provision; education systems; institutional and organizational factors (e.g., bureaucracy, corruption); cultural factors (e.g., religion, norms and values); socio-demographic factors (e.g., ethnicity, class); socio-political factors; socio-economic factors; and the physical environment.

These aspects could also be grouped under capital assets, as outlined by the British Department for International Development (DFID, 2001):

- *human capital*: knowledge, skills, abilities of individuals and people
- *social capital*: relationships, networks, collective norms, social organisation
- *natural capital*: natural resources, environmental quality, biodiversity
- *physical capital*: infrastructure, housing, energy, facilities, producer goods, communications
- *financial capital*: financial resources, available stocks, regular inflows of money.

All these aspects and assets need to be transformed through (policy, legislative, institutional and cultural) processes, which may lead to structural and sustainable change at the levels of the public and private/commercial sector and civil society. As development is an integral, multi-dimensional and dialectic process that differs from society to society, community to community, context to context, each society and community must attempt to delineate its own strategy to sustainable development starting with the resources and "capitals" available (not only physical, financial and environmental but also human, social, institutional and so on), and considering the needs and views of the people concerned.

For more overviews and more details, see McKee et al. (2000), Omoto (2005), Papa et al. (2006), Stewart et al. (2007) and Tremblay (2007). McKee et al. (2000) is a multi-sectoral and interdisciplinary synthesis of field experiences and lessons learned in the context of behavior development and change. Its aim is to challenge traditional approaches to program design, implementation and monitoring, with a view to increasing the impact and sustainability of international development programs such as the ones coordinated by the United Nations Children's Fund (UNICEF). Omoto's (2005) interdisciplinary academic volume addresses a variety of topics related to service learning, social movements, political socialization, civil society, and especially volunteerism and community development. Papa et al. (2006) address four dialectic tensions which are considered essential to the process of organizing for social change: control and emancipation, oppression and empowerment, dissemination and dialog, and fragmentation and unity. They advocate for a dialectic approach. Stewart et al. (2007) assess the contribution social networks can make to social change, while Tremblay (2007) assesses the relationship between communication and sustainable development.

We counted (see Servaes & Liu, 2007; Servaes 2007, 2008, 2012) *14 different devcom approaches,* which are currently being used and applied. Some of these are more traditional, hierarchical and linear, some are more participatory and interactive. Most contain elements of both. From an epistemological and ontological perspective, this doesn't always make sense; but in practice this seems to be a given.

In Servaes et al. (2012a, b) we designed a framework of indicators for the assessment of sustainability in CSSC projects and consequently tested them in two representative development projects. Servaes (2013) is an attempt to further elaborate on this quest for sustainability indicators, both in theory as well as in praxis.

New and ongoing challenges

Communication for sustainable development has been facing old and new challenges in the last decade, as a consequence of globalization, media liberalization, rapid economic and social change, and the emergence of new ICTs.

Right to communicate

Liberalization has led not only to greater media freedom, but also to the emergence of an increasingly consumer-led and urban-centered communication infrastructure, which is less and less interested in the concerns of the poor and rural people. Women and other vulnerable groups continue to experience marginalization, and lack of access to communication resources of all kinds. The issue of ensuring access to information and the right to communicate as a pre-condition for empowering marginalized groups has been addressed by several meetings and international conferences (World Summit on the Information Society, and the World Social Forum).

Today, Development Communication programs address global environmental issues, such as biosafety and risk management in disaster prevention and mitigation. Within this framework, communication, education, participation and public awareness approaches are used in an integrated manner to reach out effectively to the key groups who are needed to protect the environment.

Conserving environmental sustainability

Another of the challenges faced by rural development programs is how to alleviate poverty and stimulate economic growth while at the same time preserving the environment. Fighting land degradation and desertification, halting deforestation, promoting proper management of water resources and protecting biodiversity require the active participation of rural communities through communication processes.

Communication for Development focusing on participatory approaches can facilitate dialog, increase the community knowledge base (both indigenous and modern), promote agricultural practices that are compatible with the environment, and develop awareness among policy makers, authorities and service providers. Furthermore, participatory communication approaches can bring together different stakeholders and enable the poorest and most marginalized to have a voice in the use of natural resources.

Promoting food security, rural development and sustainable livelihoods

Food security and rural development policies have been revised in recent years placing more emphasis on holistic approaches to rural livelihoods and the sustainable use of natural resources. Furthermore, the spread of digital communication technology has made information and communication services an increasingly cost-effective option for providing basic information to dispersed rural producers, in particular to those settled in remote and poorly accessible areas. The focus remains on the needs of rural people, rather than on communication media *per se*. The critical aim is to enhance the capacity of local stakeholders to manage communication processes, to develop local content and to use appropriate media tools. Communication development strategies must be context specific and reflect the values, perceptions and characteristics of the people and institutions involved.

Empowerment of women, girls ... and senior citizens

Communication can also play a decisive role in promoting the empowerment of women and girls with a more equitable framework of gender dynamics. Communication processes allow rural women a voice to advocate changes in policies, attitudes and social behavior or customs. Through communication for development, women can take control of their lives and participate as equals with men in promoting food security and rural development. In a similar way, the power and expertise of senior citizens, who in many societies enjoy great authority and respect but are often left outside development efforts, should be recognized. Many societies may have to reconsider their contributions and potential anyway, given other social and demographic developments.

Narrowing the digital divide

The issue of equal access to knowledge and information is becoming one of the key aspects of sustainable development. Vulnerable groups in the rural areas of developing countries are on the wrong side of the digital divide and risk further marginalization. In the rush to "wire" developing countries, little attention has been paid to the design of ICT programs for the poor. The trend ignores many lessons learned over the years by Communication for Development approaches, which emphasize communication processes and outcomes over the application of media and technologies. There needs to be a focus on the needs of communities and the benefits of the new technologies rather than the quantity of technologies available. Local content and languages are critical to enable the poor to have access to the benefits of the information revolution. The creation of local content requires building on existing and trusted traditional communication systems and methods for collecting and sharing information. However, access is only the start of this process. Full participation implies the provision of capacity training and the development of competencies.

Good and affordable health for all

Health should be viewed from a social justice and rights-based perspective. Issues of power and gender, socio-economic determinants of the problem and both collective and individually oriented responses are equally important to tackle health problems and promote good and affordable health.

The role of the public health *analyst or researcher* is to assist, not simply a decision-maker, but a decision-making process that has the assent of the community as a whole. In this process, the *community*, as well as the decision maker and the analysts, are involved.

Poverty reduction

Communication can contribute to the effective reduction of poverty and offer better opportunities for the inclusion of marginalized groups and isolated populations in the policy-development and decision making process. Although poverty cannot be divorced from uneven power structures, and communication cannot be a substitute for structural change, the appropriation of communication for development processes and technologies by marginalized and vulnerable groups, including indigenous people, can ensure that they have a voice in decisions that affect their lives.

The way forward

This brief attempt to outline future challenges for the field, indicates that CSSC&D approaches differ according to what development issues are involved. What they have in common is a set of guiding principles and steps to follow. The emphasis now is on the *process of communication* and on the *significance of this process at the local level*. Furthermore, according to the approach of different agencies, CSSC coincides with the enhancement of local capacities and the appropriation of communication processes and media by local stakeholders, and especially by vulnerable and marginalized groups. Therefore, *capacity building* in communication, including "bridging" of the digital divide is now seen as an essential condition for sustainable development and the fulfilment of the United Nations Millennium Development Goals; soon to be renamed as *Sustainable Development Goals* (SDGs).

Three streams of action are important:

- *Media* must be activated to build public support and upward pressure for policy decisions.
- *Interest groups* must be involved and alliances established for reaching a common understanding and mobilizing societal forces. This calls for networking with influential individuals and groups, political forces and public organizations, professional and academic institutions, religious and cause-oriented groups, business and industry.

- *Public demand* must be generated and citizens' movements activated to evoke a response from local, national and international leaders. Organized social action with the involvement of committed individuals and communities, support from influential forces and the involvement of concerned sectors of society will result in (more) power to the people.

The 2013 Human Development Report concludes that

> the rise of the South demonstrates that the world has become more diffuse and cross-connected. One consequence is that rather than looking to the North for inspiration, developing countries are looking to their peers in the South for appropriate development models. Here, rather than seeing a sterile menu of ideological options, they can examine what has worked, under what circumstances, and choose the most appropriate tools.
>
> (Malik, 2013: 119)

We sincerely hope that our book has made a modest contribution in this regard toward a better understanding of the place and role of communication for sustainable development and social change.

References

Dfid (2001). *Sustainable Livelihoods Guidance Sheets, Section 2: Framework*. London: Department of International Development, 26pp, www.livelihoods.org/onfo/info_guidanceSheets.html

McKee, N., Manoncourt, E., Saik Yoon, C., & Carnegie, R. (2000). *Involving People Evolving Behaviour*. Penang: Southbound.

Malik, K. (ed.) (2013). *The Rise of the South: Human Progress in a Diverse World. Human Development Report*. New York: UNDP.

Omoto, A. (2005). *Processes of Community Change and Social Action*. Mahwah, NJ: Lawrence Erlbaum.

Papa, M.J., Singhal, A., & Papa, W. 2006. *Organizing for Social Change: A Dialectic Journey of Theory and Praxis*. New Delhi: Sage Publications.

Servaes, J. (2007). Harnessing the UN System into a Common Approach on Communication for Development, *The International Communication Gazette*, 69(6), pp. 483–507.

Servaes, J. & Liu, S. (eds.) (2007). *Moving Targets: Mapping the Paths Between Communication, Technology and Social Change in Communities*. Penang: Southbound.

Servaes, J. (ed.) (2008). *Communication for Development and Social Change*. London: Sage.

Servaes, Jan (2012) Comparing Development Communication. In Frank Esser & Thomas Hanitzsch (eds.) *The Handbook of Comparative Communication Research*, Routledge: New York, pp. 64–80.

Servaes, Jan, Emily, P., Shi, S. Reilly, D., & Yakupitijage, T. (2012a), Towards a Framework of Sustainability Indicators for "Communication for Development and Social Change" Projects, *The International Communication Gazette*, 74(2), pp. 99–123.

Servaes, J., Emily, P., Shi, S., Reilly, D., & Yakupitijage, T. (2012b), Sustainability Testing for Development Projects, *Development in Practice*, 22(1), February, pp. 18–30.

Servaes, Jan (ed.) (2013), *Sustainability, Participation and Culture in Communication: Theory and Praxis*. Bristol/Chicago: Intellect/University of Chicago Press.

Stewart, C., Smith, C. A., & Denton, R. (2007). *Persuasion and Social Movements* (5th ed.). Long Grove, IL: Waveland Press.

Tremblay, S. (Ed.) (2007). *Developpement durable et communications. Au-dela des mots, pour un veritable engagement*. Quebec: Presses de l'Universite du Quebec.

Index